PULMONARY NANOMEDICINE

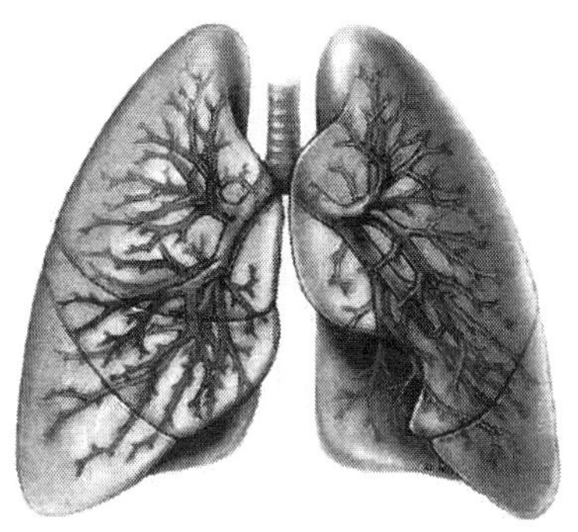

EDITED BY

NEERAJ VIJ

PULMONARY NANOMEDICINE

DIAGNOSTICS, IMAGING, AND THERAPEUTICS

Published by

Pan Stanford Publishing Pte. Ltd.
Penthouse Level, Suntec Tower 3
8 Temasek Boulevard
Singapore 038988

Email: editorial@panstanford.com
Web: www.panstanford.com

British Library Cataloguing-in-Publication Data
A catalogue record for this book is available from the British Library.

Pulmonary Nanomedicine: Diagnostics, Imaging, and Therapeutics

ISBN 978-981-4316-48-4 (Hardcover)
ISBN 978-981-4364-14-0 (eBook)

Printed in the USA

To the fond memory of my parents,
who taught me to seek enlightenment and knowledge
and strive for excellence.

Dearest souls,
although you have departed and moved far away,
your perpetual memory in our hearts
makes us feel that you are ever near us, with us.

—N.V.

Contents

Contributors

James C. Bonner received his PhD in physiology from Mississippi State University in 1987, completed his postdoctoral training at the National Institute of Environmental Health Sciences (NIEHS) in 1990, and served as a principal investigator at NIEHS and at the Hamner Institutes for Health Sciences. He joined the Department of Environmental and Molecular Toxicology at NC State University as an associate professor in 2007.

Dr. Bonner has over 20 years of experience in respiratory toxicology and lung disease pathogenesis. He has published more than 80 peer-reviewed research articles on environmental lung disease, numerous review articles, and several textbook chapters on respiratory toxicology.

He is the lead author of Chapter 9, on the potential respiratory health risks of engineered carbon nanotubes.

Aaron E. Foster received his BA in biology from the University of Puget Sound in Tacoma, Washington, in 1994 and his PhD in chemical engineering from the University of Sydney, Australia, in 2003. Currently, he is an assistant professor at the Center for Cell and Gene Therapy at Baylor College of Medicine.

Dr. Foster's research interests include cancer vaccine development, immune modulation, and gene therapy applications using cytotoxic T lymphocytes (CTLs) as anti-tumor effector cells or as carriers for *in vivo* delivery. In collaboration with Dr. Rebekah Drezek in the Department of Bioengineering at Rice University, he is also studying the development and use of multifunctional nanoparticles for the treatment of cancer and infectious disease.

Dr. Foster is the main author of Chapter 5, on the use of nanoparticles to treat airway inflammation.

Song Li received his MD in 1985 and PhD in tumor immunotherapy in 1991 from the Fourth Military Medical University, China. He worked with Dr. Leaf Huang at the University of Pittsburgh School of Medicine as a post-doc for two years and then as a research faculty for another four years. Dr. Li joined the faculty of the School of Pharmacy at the University of Pittsburgh in June 2000 and is currently an associate professor of pharmaceutical sciences. His major research interest is focused on the development of lipid- and polymer-based nanodelivery systems for targeted delivery of various types of therapeutics including nucleic acids (genes, siRNA, and peptide nucleic acids), proteins, and small molecules (e.g., anticancer agents and antioxidants).

Dr. Li is the main author of Chapter 7, on targeted delivery to the pulmonary endothelium.

Ashim K. Mitra received his PhD in pharmaceutical chemistry in 1983 from University of Kansas. He joined the University of Missouri–Kansas City in 1994 as chairman of Pharmaceutical Sciences. He is also vice provost for Interdisciplinary Research, Curators' Professor of Pharmacy, and director for Translational Research at University of Missouri–Kansas City, School of Medicine. Dr. Mitra has over 25 years of experience in ocular drug delivery and disposition and has authored or co-authored over 250 refereed articles and 30 book chapters in this field. He holds 8 patents and has made well over 450 presentations, including abstracts, at national and international scientific meetings. This work has attracted over 6 million dollars in funding from government agencies such as the National Institutes of Health (NIH) and Department of Defense (DOD) and from pharmaceutical companies.

Dr. Mitra is also a recipient of several research awards from AAPS, AACP, and various pharmaceutical organizations and serves on numerous editorial boards. According to Biomed Experts (during the past 10 years), he co-authored the third-highest number of publications in the world in the area of "Prodrugs." In April 2010, he was ranked fifth in the world among AAPS's Top Ten Researchers. In February 2012, his article "Ocular Drug Delivery" was again ranked as one of the top 5 downloaded articles in the *AAPS Journal*. Currently, he is chairman of the USP Council of Experts, General Chapter

<771> *Ophthalmic Preparations* Expert Panel, U.S. Pharmacopeia. His current research interests are focused on two main areas: delivery and targeting of antiviral agents and development of noninvasive delivery systems for peptide and protein drugs.

Dr. Mitra is the senior author of Chapter 3, which discusses the use of nasal and pulmonary delivery of macromolecules to treat respiratory and nonrespiratory diseases.

Shyam S. Mohapatra is a Distinguished USF Health Professor and director of the Division of Translational Medicine, Nanomedicine Research Center at the Morsani College of Medicine, University of South Florida. He also directs the Signature Program in Allergy, Immunology and Infectious Diseases at the college. A PhD graduate of the Australian National University, Prof. Mohapatra is a recipient of two international awards: the Alexander von Humboldt research fellowship (1984, Bonn, Germany) in genetics and Pharmacia Allergy Research Foundation Award (1992, Paris) for his contributions to the field of allergy and immunology.

Prof. Mohapatra's research program focuses on the molecular mechanisms underlying inflammation in respiratory diseases, cancers, viral infections, and traumatic brain injury. He has used nanotechnology approaches to advance translational research in these disease areas.

Prof. Mohapatra is the senior author of Chapter 6, which discusses the application of multifunctional chitosan nanocarriers in respiratory gene therapy.

Rajagopal Ramesh received his PhD in molecular biology in 1994 from the All India Institute of Medical Sciences, New Delhi, India. He completed his postdoctoral fellowships at Tulane University School of Medicine, New Orleans, in 1998 and later joined the faculty at M. D. Anderson Cancer Center in Houston, USA. Currently, Dr. Ramesh is a professor in the Department of Pathology and director of Experimental Therapeutics and Translational Cancer Medicine at the University of Oklahoma Health Sciences Center, Oklahoma City, OK, USA. He holds the Jim and Christy Everest Endowed Chair in Cancer Developmental Therapeutics and the title of the Oklahoma TSET Cancer Research Scholar.

Dr. Ramesh's research is focused on investigating new cancer therapies with an emphasis on lung cancer. His laboratory has been conducting applied translational cancer research in leading areas such as cancer gene therapy, nanotechonology, and molecular targeted therapy. Research in his laboratory has led to four clinical trials for the treatment of cancer. He has published more than 86 articles in leading scientific journals and 17 textbook chapters on cancer therapy and drug development. His research is funded by the National Cancer Institute and other national granting agencies.

Dr. Ramesh is the senior author of Chapter 2, which discusses the use of multifunctional tumor-targeted nanoparticles for lung cancer.

Indrajit Roy received his MSc and PhD degrees from the University of Delhi, India, in 1997 and 2002, respectively. Following that, he completed postdoctoral research at the State University of New York (SUNY) at Buffalo, as well as at the Johns Hopkins School of Medicine. His research interests include the use of various nanoparticles for applications in targeted drug delivery, nonviral gene delivery, photodynamic therapy (PDT), and multimodal diagnostic imaging.

Dr. Roy has published more than 50 articles in leading scientific journals and holds 3 U.S. patents. In 2005, he was presented with the Visionary Innovator Award by the technology transfer office at SUNY, Buffalo. From 2005 to 2009, he served as a research assistant professor in the Institute for Lasers, Photonics and Biophotonics (ILPB) at SUNY, Buffalo. At present, he is an associate professor of chemistry at the University of Delhi.

Prof. Roy is the author of Chapter 4, on *in vitro* and *in vivo* diagnosis of pulmonary disorders using nanotechnology.

Venkataramana K. Sidhaye received her bachelor's in biomedical engineering in 1995 and her MD in 1998 at Northwestern University. She then did her residency in internal medicine and was a chief resident at Northwestern before coming to Johns Hopkins in 2002 for fellowship training in pulmonary and critical care medicine. She joined the Hopkins Pulmonary faculty in 2006. Dr. Sidhaye's research interest in is epithelial barrier function, with a focus on the airway epithelium. She is interested in the cross-talk between the epithelial barrier and cell–cell contacts and the role of the epithelium in innate immunity, and this is modified by luminal exposures. More recently, she has been interested in epithelial responses to inspired nanomaterials.

Dr. Sidhaye is the senior and corresponding author of Chapter 8, on epithelial barrier targeting in chronic airway diseases.

Neeraj Vij received his PhD in biotechnology from the Indian Institute of Technology in 2001 and was also a recipient of international fellowship at the Institute of Genetics, Biological Research Center (Centre of Excellence of the European Union), Hungary, in same year. He subsequently completed his postdoctoral research at the University of Heidelberg, Germany, and The Johns Hopkins University School of Medicine (JHU SOM). Dr. Vij is currently an assistant professor at the Department of Pediatric Respiratory Sciences and Institute of NanoBiotechnology, JHU SOM. He serves on the editorial boards of several nanomedicine journals, including *Journal of Nanomedicine & Nanotechnology, Expert Opinion in Drug Delivery, International Journal of Nano Studies & Technologies*, and so forth. He has been invited to help organize several nanotechnology conferences and seminars, such as NanoBiotech 2009, and the nanotechnology postgraduate course at American Thoracic Society (ATS). He is also frequently invited to serve as a reviewer for various nanomedicine journals and grant review study sections, including NIH, USA. He is a life member of the American Society for Nano Medicine (ASNM) and several other international scientific societies. Dr. Vij has received several research awards and recognition for his scientific contributions.

The primary research focus of Dr. Vij's laboratory is identification of molecular pathways leading to chronic disease pathophysiology, with an aim to identify novel therapeutic sites. His laboratory is interested in applied and pre-clinical translational research and concentrates on the identification of novel therapeutic strategies including design and development of nano-based delivery systems for theranostic applications in chronic obstructive lung diseases.

Dr. Vij is the editor of this book and senior author of Chapter 1, which discusses the theranostic applications of nanotechnology in chronic obstructive lung diseases.

Preface

Nanotechnology has revolutionized medicine over the past decade. The unique physicochemical characteristics of engineered nanoparticles (ENPs) enable novel therapeutic and diagnostic (theranostic) applications, particularly in pulmonary diseases. The research over the past decade has provided insights into biological properties and application of NPs in pulmonary medicine.

This book provides a comprehensive review on the pulmonary applications of NPs and aims to enlighten the readers about novel nano-based theranostic strategies for treating pulmonary disorders. Each chapter discusses strategies to overcome the technological and disease-specific pathophysiological barriers to develop novel nano-based diagnostics, imaging, and therapeutic tools for treatment of various airway diseases.

In summary, the book is focused on emerging cutting-edge applications of nanotechnology in pulmonary medicine and aims to synchronize the efforts of pulmonary biologists, nano-chemists, and clinicians to develop novel nano-based theranostic systems for treatment of airway diseases.

This book has been compiled with the goal to serve both academic institutions and industry for education, training, and research. It is written to educate graduate and postgraduate students on emerging theranostic applications of ENPs in treating various pulmonary diseases. It will also serve as a guide for both clinicians and researchers in developing novel theranostics while closely monitoring the health effects of next-generation ENPs.

Overall, this is a wikipidea of pulmonary nanomedicine that discusses the scope of both current and future nanotechnologies for pulmonary applications.

Neeraj Vij, MS, PhD
Baltimore, MD
April 2012

Acknowledgments

I express my sincere thanks to all authors and reviewers, who are experts in their respective fields, for their exceptional contribution and support. This book came into its present form with the earnest efforts of all authors, who helped me ensure that therapeutic and diagnostic strategies of novel pulmonary nanomedicine were discussed to help lead the advancement of the emerging scientific field of pulmonary nanomedicine. I am extremely grateful to Stanford Chong (Director) and Sarabjeet Garcha (Editorial Manager), of Pan Stanford Publishing Pte. Ltd., for their outstanding support and perceptiveness. Sarabjeet was especially instrumental in providing the much-needed editorial support for the swift collation of the book.

Neeraj Vij

Chapter 1

Theranostic Applications of Nanotechnology in Chronic Obstructive Lung Diseases

Neeraj Vij* and Aakruti Gorde
Department of Pediatric Respiratory Sciences and Institute of NanoBioTechnology, Johns Hopkins School of Medicine, Baltimore, MD, USA
*nvij1@jhmi.edu

1.1 Pulmonary Physiology and Pathogenesis of Chronic Obstructive Lung Diseases

The respiratory tract can be divided into two parts — the upper respiratory tract, which includes the nose, nasal cavity, pharynx, epiglottis, and larynx, and the lower respiratory tract, which includes the trachea, bronchi, and lungs. Lungs are primarily responsible for oxygen delivery to the cells of the body, removing carbon dioxide by exhalation, and maintenance of homeostatic systemic pH. Beginning at the end of the larynx, the trachea divides into two bronchi, which further branch into smaller bronchioles. The terminal parts of the bronchi end up in the alveolar sac. The conducting airways are lined with ciliated columnar epithelium, which transition to a cuboidal shape approaching the distal airways. The lumen of the

Pulmonary Nanomedicine: Diagnostics, Imaging, and Therapeutics
Edited by Neeraj Vij

ISBN 978-981-4316-48-4 (Hardback), 978-981-4364-14-0 (eBook)
www.panstanford.com

bronchial airways is lined by a thin layer of serous fluid covered with a layer of mucus, which helps to entrap aerosolized particles [http://en.wikibooks.org/wiki/Human_Physiology/The_respiratory_system] [1].

The anatomy and physiology of the human respiratory tract was recently reviewed in an article covering the application of nanomaterials in respiratory disorders [2]. Briefly, the alveoli form a net of cells around the spiral cylindrical surface of the alveolar duct. The alveolar surface is primarily composed of type I pneumocytes, which are nonphagocytic and surface-line the membranous epithelial cells. Attached to the basement membranes are the larger type II pneumocytes, which secrete a lung surfactant to prevent an alveolar collapse. Although alveoli are tiny structures, there are approximately 300 million of them in each lung, with a total surface area greater than 100m^2. This large surface area, combined with the extremely thin alveolar epithelium between the pulmonary lumen and capillaries, is responsible for efficient mass transfer [2, 3]. Moreover, inflammatory cells such as alveolar macrophages ensure the clearing of large nano- or foreign particles.

Our respiratory system is constantly exposed to harmful environmental pollutants throughout our lifetime, which makes it highly susceptible to chronic insults. Although our lungs are equipped with robust defensive mechanisms, they sometimes succumb to the continuous deleterious effects of environmental pollutants. Long-term exposure to these detrimental agents collectively leads to chronic obstructive pulmonary disease (COPD) [4, 5]. COPD is an umbrella term for a group of conditions in which there is persistent difficulty in exhaling air from the lungs. COPD commonly refers to two related, progressive diseases of the respiratory system, chronic bronchitis, and emphysema [4, 5]. The major cause of COPD is cigarette smoking, which causes chronic inflammation, leading to the narrowing of the bronchi, which interferes with the flow of air [4–6]. Moreover, chronic inflammation causes excessive mucus secretion, which further blocks the airways [5]. Cigarette smoke (CS) also damages the cilia in the respiratory tract, increasing the risk of chronic infections [5, 7]. The most damaging effect of CS is the development of emphysema, wherein there is irreversible destruction of the lung alveoli, the terminal air sacs of the respiratory system [5, 8]. The pathophysiology of COPD involves persistent inflammation, oxidative stress, impaired lung cell repair, and programmed cell death (apoptosis) leading to emphysema [8, 9]. Both chronic bronchitis and emphysema can lead to respiratory failure, which can ultimately have fatal consequences. Moreover, individuals with a genetic deficiency of α1-AT can also develop COPD-like disease that involves emphysema and chronic bronchitis [10]. In a small percentage of COPD subjects, a pathogenic mutation in the α1-AT gene results in misfolding of α1-AT protein (Z-variant, ATZ), which is degraded via the proteasomal pathway, leading to a deficiency of the functional protein [10]. Similarly, in case of a common protein-folding mutation that causes cystic fibrosis (CF) lung disease, the deletion of phenylalanine at position 508 of CF transmembrane conductance regulator gene (ΔF508-CFTR) results in a temperature-sensitive folding defect and premature degradation of CFTR protein by endoplasmic reticulum–associated degradation (ERAD) [11–14], leading to the pathogenesis of fatal obstructive lung disease [15]. In summary, both environmental and genetic factors promote the pathogenesis of chronic obstructive disease, which can be further aggravated by infection and age [16, 17].

1.2 Application of Nano-Based Systems in Treating Chronic Obstructive Lung Diseases

We and others have recently discussed in detail the application of novel nanosystems in treating various respiratory conditions such as CF, COPD, allergy, asthma, and lung cancer [18–27]. We anticipate that stringent preclinical evaluation and standardization of novel disease-specific nanotherapeutic strategies hold a promise for clinical application and translation. Drug or gene delivery via nanosystems face various disease-related pathophysiological challenges. Hence, nanosystems specifically designed to circumvent these challenges in the specific disease state need to be carefully designed [18, 19].

On airway deposition, nanoparticles encounter various physiochemical and biological barriers such as the catabolic enzymes and mucus in the tracheobronchial region and macrophages in the alveolar region. In the peripheral region of the lung, particles must be able to dissolve and diffuse through the epithelial barrier into the blood stream. Large-size particles that are unable to do so are subject to phagocytosis by alveolar macrophages. However, this property of large particles can be tailored to selectively target the drug to macrophages in disease states such as COPD-emphysema [19]. On the contrary, small- or ultrafine-nanoparticles have the tendency to accumulate in the airway. The surface coating on the nanosystems can be used to avoid their aggregation while promoting clearance. Despite these hindrances, the lung is an attractive target for nanoparticle-mediated drug delivery due to its capacity to provide a noninvasive means for local lung delivery as well as high systemic bioavailability. A large surface area for absorption and limited proteolytic activity makes the pulmonary route an excellent system for local delivery of drugs for the treatment of various obstructive lung diseases, such as asthma, COPD-emphysema, CF, and as other respiratory conditions, including pulmonary hypertension and lung cancer. It is advantageous to treat these diseases locally since the drug is able to directly deposit itself at the disease site, avoiding rapid metabolism. As an example, nanoparticle-mediated drug delivery to the lung epithelium eliminates potential side effects caused by high systemic concentrations and reduces costs by use of both small doses and targeted drug activity [28]. Moreover, the nano-based systems are ideal for targeted delivery of drugs to specific inflammatory cells, such as alveolar macrophages, neutrophils, or T cells, for the treatment of obstructive lung diseases, as we recently discussed [19].

Obstructive lung diseases such as COPD represent one of the major global causes of disability and death. It is estimated that COPD will become the third-leading cause of death by 2020 [5]. As we recently discussed in detail, the major challenges in the delivery and therapeutic efficacy of nanodelivery systems in chronic obstructive lung diseases are severe inflammation and mucous hypersecretion [19–23], which are exacerbated by infection or components of CS [21, 22, 24–26]. The chronic stage of these diseases is associated with widespread damage to the airway cells due to excessive inflammation, apoptosis and defective repair mechanisms [22, 24, 26, 27]. Controlling chronic inflammatory-oxidative responses is an acceptable or available therapy, although the challenge is the targeted and controlled delivery of drugs [28]. We have also recently discussed in detail that in spite of the wide application of nano-based drug delivery systems in chronic obstructive airway diseases and a variety of other pulmonary conditions, very few are tested till date [5, 7, 29, 30]. As an example, we have recently discussed the application of epithelial-targeted nanoparticles to control neutrophil chemotaxis, fibrosis, and protease-mediated chronic emphysema (Fig. 1.1) [19].

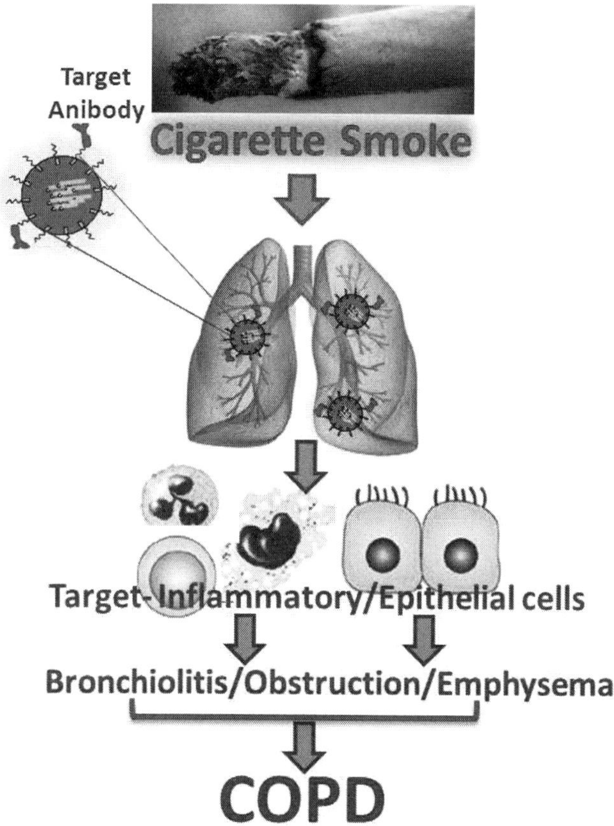

Figure 1.1 Application of nanosystems in the treatment of COPD-emphysema. Adapted from Vij [19]. See also Color Insert.

As discussed above, in monogenetic lung disorders, such as CF and α1-AT deficiency, gene replacement therapy is an ideal therapeutic strategy but chronic inflammatory response and obstructive lung disease pose a challenge for efficient and stable nanoparticle-mediated gene delivery. In addition, nanoparticles also provide an ideal drug delivery system for the epithelial-targeted release of CFTR correctors/potentiators and anti-inflammatory drugs [20].

1.2.1 Therapeutic and Diagnostic Challenges in Chronic Obstructive Lung Diseases

As discussed above and reviewed recently in detail [18, 19], a major challenge in the delivery and therapeutic efficacy of nanodelivery systems in chronic obstructive lung diseases is severe inflammation and mucous hypersecretion [18–20, 29]. Distinct mechanisms drive these responses in these disease states. The chronic inflammatory response is macrophage (COPD) or neutrophilic (CF and COPD) driven [5, 30–33] and may be induced by infection, injury or CS. The chronic stage of these diseases is associated with severe injury to the bronchial epithelium due to excessive inflammation and apoptosis and defective epithelial repair. Controlling chronic inflammatory-oxidative responses is an acceptable or available therapy, although the challenge is the targeted and controlled delivery of drugs [18]. As we recently discussed in detail, nano-based drug delivery systems can override these challenges in chronic obstructive airway diseases and other pulmonary conditions but very few are tested clinically till date [18, 19, 34]. The chronic obstructive lung diseases also pose similar challenges for real-time assessment of the lung disease by molecular imaging. The efficient molecular probe delivery is critical for real-time assessment of the lung diseases. We have also recently discussed that nanosystem-mediated molecular probe delivery can provide an optimal tool for both the diagnosis and assessment of therapeutic efficacy [20]. Moreover, a nanosystem with imaging capability can provide image-guided targeted drug delivery [35], but the challenge in obstructive lung disease is to bypass the inflammatory and obstructive barrier to achieve targeted delivery to the diseased cells.

1.2.2 Nanosystems to Overcome Challenges in Chronic Obstructive Lung Diseases

We have discussed in detail the efficacy and potential application of PLGA-based nanosystems for drug delivery in chronic obstructive lung diseases [36, 37]. Polymeric nanoparticles exhibit several desirable properties as gene, drug, and/or molecular probe carriers, including properties such as biocompatibility, biodegradability, and functionalization capability. Using an appropriate method, a gene, therapeutic drug, and/or molecule probe can be attached to the polymer/copolymer for controlled release at the disease site. Techniques that are commonly used to load the theranostic in a nanosystem such as a polymer includes impounding, conjugation, and micelle formation [18, 38].

Polymeric nanoparticle structures can be synthesized by various methods that are recently summarized [38]. Moreover, several recent reviews have discussed the therapeutic/gene delivery potential of various biodegradable polymer-based nanomaterials. We reviewed their applications in treating chronic obstructive lung diseases. Biodegradable and biocompatible polymers are promising candidates for theranostic applications [19], although some of these polymer, such as poly-(lactic-co-glycolic acid) (PLGA) is highly hydrophobic and have a high negative charge on their surface. As a result, they are rapidly opsonized by the airway defense system, as discussed earlier [18]. The commonly used method to overcome these challenges is to coat the nanosystem with a copolymer such as polyethylene glycol (PEG), which not only protects the nanosystem from immediate clearance but also provides a method for conjugating antibodies for cell-specific targeted drug delivery (Fig. 1.1). We have recently reviewed that PEGylation also increases the circulation time of the nanoparticles, thereby enhancing their probability of accumulation in target organs or cells by passive diffusion involving the enhanced permeability and retention (EPR) effect [18]. These $PLGA^{PEG}$ nanoparticles also increase the residence time of the nanoparticle cargo in blood (intravenous) or airway (intranasal), providing sustained and targeted drug/gene delivery to specific cells or tissues [18]. As discussed previously, the major challenges in the delivery and theranostic efficacy of nanodelivery systems in chronic obstructive airway conditions are airway defense (such as macrophage and neutrophils), severe inflammation, and mucous hypersecretion [19].

1.3 Nanotheranostics

Theranostics is the pairing of a diagnostic test with a therapy (therapeutic) and is considered the pathway to personalized medicine that will usher in an entirely new era of healthcare delivery

[35, 39–43]. Nano-based theranostics can offer unique applications such as real-time monitoring of drug delivery systems and disease states. Moreover, it provides assessment of the efficacy of treatment for optimal therapy of individual patients and reduction in adverse drug effects [19, 44]. Several multifunctional nanoparticles have been reported for treatment of various forms of cancer [45–48]. Nano-based delivery systems offer a number of advantages over conventional drug delivery systems. To make theranostics possible for pulmonary disease conditions such as lung cancer, asthma, allergy, infection, and inflammation, it is crucial to develop multifunctional airway-targeted nanocarriers for combined delivery of diagnostic and therapeutic agents, as we recently discussed [19]. These multivalent nano-based systems allow the integration of multiple functions, such as cell-specific targeting, and cargo for imaging modality and therapeutic.

Emerging nano-theranostics [35, 44] are a promising treatment for chronic obstructive lung diseases as they provide cell-targeted therapy with the ability to quantify the efficacy of the treatment or changes in the disease state. As we and others have recently discussed, these nano-theranostic carriers can be used to improve the theranostic efficacy of new or established drugs/molecular probes by providing an enhanced site-directed therapeutic effect [19, 35]. The theranostic efficiency of these nano-based systems is dependent upon both composition and physical-chemical properties. An efficient nano-theranostic system provides drug/molecular probe delivery to target tissue/cells and provides image-guided drug delivery for achieving optimal theranostic activity [19, 35].

1.3.1 Theranostic Nanoparticles for Chronic Obstructive Lung Diseases

We have recently discussed in detail the use of multifunctional polymeric vesicles such as PLGA^{PEG} [18, 20] for the combined delivery of drugs and/or molecular probes in chronic obstructive lung diseases [19]. We proposed that it is essential to develop multifunctional airway-targeting nanocarriers for the combined delivery of diagnostic and therapeutic agents as it may greatly help in correcting or reverting these diseases from chronic states. These nanotheranostic systems may not only allow targeted drug/gene delivery to disease cell type but also provide imaging ultra-sensitivity for real-time assessment of the disease state and therapeutic efficacy. A variety of multifunctional nanoparticles have been reported for treating various types of cancers [49, 50]. As we recently discussed, these multifunctional nanosystems are under different stages of development for cancer therapy [51–59] and diagnosis [45, 60, 61] but they need to be carefully evaluated before clinical applications [62]. Since most of these nanotheranostic materials are quantum dots, metals, or metal oxides, tumor-targeted delivery of these nanosystem is essential to avoid toxicity for nontumor cells. Moreover, the application of these systems in chronic obstructive lung diseases or other respiratory disorders is limited as they pose the risk of further aggravating the lung disease by inducing inflammatory-oxidative stress and toxicity. Alternatively, we discussed [19] the use of amphiphilic block copolymers that can form several types of nano-assemblies in an aqueous solution, including micelles and vesicles, depending on their relative hydrophilicity/hydrophobicity and chemical structures [63, 64]. Moreover, compared with liposomes, polymer-based vesicles offer numerous possibilities for controlling the physiochemical and biological properties of nanoparticles in order to tailor their shape, size, and functionalization properties [64, 65].

1.3.2 Perspective

Nanotheranostic systems are considered to be optimal systems for both preclinical experimental evaluation and clinical implementation as they can provide real-time assessment of therapeutic delivery and efficacy while monitoring changes in the disease state [19, 35]. However, nanotheranostic systems require the identification of highly specific molecular probes for the detection of specific inflammatory cells or response, mucus-obstruction, and/or injury. Nanotheranostics is an emerging field that requires the identification and stringent evaluation of novel biodegradable nanomaterials for application in chronic obstructive lung disease, as discussed in this chapter.

References

1. Bailey MM, Gorman EM, Munson EJ, Berkland C: Pure insulin nanoparticle agglomerates for pulmonary delivery. *Langmuir* 2008, 24:13614–13620.
2. Smola M, Vandamme T, Sokolowski A: Nanocarriers as pulmonary drug delivery systems to treat and to diagnose respiratory and non respiratory diseases. *Int J Nanomedicine* 2008, 3:1–19.
3. Burns AR, Smith CW, Walker DC: Unique structural features that influence neutrophil emigration into the lung. *Physiol Rev* 2003, 83:309–336.
4. Barnes PJ: New concepts in chronic obstructive pulmonary disease. *Annu Rev Med* 2003, 54:113–129.
5. Barnes PJ, Shapiro SD, Pauwels RA: Chronic obstructive pulmonary disease: molecular and cellular mechanisms. *Eur Respir J* 2003, 22:672–688.
6. Rahman I, Adcock IM: Oxidative stress and redox regulation of lung inflammation in COPD. *Eur Respir J* 2006, 28:219–242.
7. Cantin AM: Cellular response to cigarette smoke and oxidants: adapting to survive. *Proc Am Thorac Soc* 2010, 7:368–375.
8. Yoshida T, Tuder RM: Pathobiology of cigarette smoke-induced chronic obstructive pulmonary disease. *Physiol Rev* 2007, 87:1047–1082.
9. Filosto S, Castillo S, Danielson A, Franzi L, Khan E, Kenyon N, Last J, Pinkerton K, Tuder R, Goldkorn T: Neutral sphingomyelinase 2: a novel target in cigarette smoke-induced apoptosis and lung injury. *Am J Respir Cell Mol Biol* 2011, 44:350–360.
10. Gooptu B, Ekeowa UI, Lomas DA: Mechanisms of emphysema in alpha1-antitrypsin deficiency: molecular and cellular insights. *Eur Respir J* 2009, 34:475–488.
11. Vij N: AAA ATPase p97/VCP: cellular functions, disease and therapeutic potential. *J Cell Mol Med* 2008, 12:2511–2518.
12. Kopito RR: ER quality control: the cytoplasmic connection. *Cell* 1997, 88:427–430.
13. Kopito RR: Biosynthesis and degradation of CFTR. *Physiol Rev* 1999, 79:S167–S173.
14. Kopito RR: Aggresomes, inclusion bodies and protein aggregation. *Trends Cell Biol* 2000, 10:524–530.
15. Bodas M, Vij N: The NF-kappaB signaling in cystic fibrosis lung disease: pathophysiology and therapeutic potential. *Discov Med* 2010, 9:346–356.

16. Bodas M, Min T, Vij N: Early-age-related changes in proteostasis augment immunopathogenesis of sepsis and acute lung injury. *PLoS One* 2010, 5:e15480.

17. Ito K, Barnes PJ: COPD as a disease of accelerated lung aging. *Chest* 2009, 135:173–180.

18. Roy I, Vij N: Nanodelivery in airway diseases: challenges and therapeutic applications. *Nanomedicine* 2010, 6:237–244.

19. Vij N: Nano-based theranostics for chronic obstructive lung diseases: challenges and therapeutic potential. *Expert Opin Drug Deliv* 2011, 8:1105–1109.

20. Vij N, Min T, Marasigan R, Belcher CN, Mazur S, Ding H, Yong KT, Roy I: Development of PEGylated PLGA nanoparticle for controlled and sustained drug delivery in cystic fibrosis. *J Nanobiotechnology* 2010, 8:22.

21. John AE, Lukacs NW, Berlin AA, Palecanda A, Bargatze RF, Stoolman LM, Nagy JO: Discovery of a potent nanoparticle P-selectin antagonist with anti-inflammatory effects in allergic airway disease. *Faseb J* 2003, 17:2296–2298.

22. Kong X, Zhang W, Lockey RF, Auais A, Piedimonte G, Mohapatra SS: Respiratory syncytial virus infection in Fischer 344 rats is attenuated by short interfering RNA against the RSV-NS1 gene. *Genet Vaccines Ther* 2007, 5:4.

23. Kumar M, Kong X, Behera AK, Hellermann GR, Lockey RF, Mohapatra SS: Chitosan IFN-gamma-pDNA Nanoparticle (CIN) Therapy for Allergic Asthma. *Genet Vaccines Ther* 2003, 1:3.

24. Oyarzun-Ampuero FA, Brea J, Loza MI, Torres D, Alonso MJ: Chitosan-hyaluronic acid nanoparticles loaded with heparin for the treatment of asthma. *Int J Pharm* 2009, 381:122–129.

25. Gopalan B, Ito I, Branch CD, Stephens C, Roth JA, Ramesh R: Nanoparticle based systemic gene therapy for lung cancer: molecular mechanisms and strategies to suppress nanoparticle-mediated inflammatory response. *Technol Cancer Res Treat* 2004, 3:647–657.

26. Reynolds C, Barrera D, Jotte R, Spira AI, Weissman C, Boehm KA, Pritchard S, Asmar L: Phase II trial of nanoparticle albumin-bound paclitaxel, carboplatin, and bevacizumab in first-line patients with advanced nonsquamous non-small cell lung cancer. *J Thorac Oncol* 2009, 4:1537–1543.

27. Sun W, Fang N, Trewyn BG, Tsunoda M, Slowing, II, Lin VS, Yeung ES: Endocytosis of a single mesoporous silica nanoparticle into a human lung cancer cell observed by differential interference contrast microscopy. *Anal Bioanal Chem* 2008, 391:2119–2125.

28. Ulbrich W, Lamprecht A: Targeted drug-delivery approaches by nanoparticulate carriers in the therapy of inflammatory diseases. *J R Soc Interface* 2010, 7 Suppl 1:S55–S66.

29. Wine JJ: The genesis of cystic fibrosis lung disease. *J Clin Invest* 1999, 103:309–312.

30. Barnes PJ: Mediators of chronic obstructive pulmonary disease. *Pharmacol Rev* 2004, 56:515–548.

31. Barnes PJ: The cytokine network in asthma and chronic obstructive pulmonary disease. *J Clin Invest* 2008, 118:3546–3556.

32. Ulrich M, Worlitzsch D, Viglio S, Siegmann N, Ladarola P, Shute JK, Geiser M, Pier GB, Friedel G, Barr ML, *et al.*: Alveolar inflammation in cystic fibrosis. *J Cyst Fibros* 2010, 9:217–227.

33. Caito S, Rajendrasozhan S, Cook S, Chung S, Yao H, Friedman AE, Brookes PS, Rahman I: SIRT1 is a redox-sensitive deacetylase that is post-translationally modified by oxidants and carbonyl stress. *Faseb J* 2010, 24:3145–3159.

34. Yang W, Peters JI, Williams RO, 3rd: Inhaled nanoparticles--a current review. *Int J Pharm* 2008, 356:239–247.

35. Lammers T, Kiessling F, Hennink WE, Storm G: Nanotheranostics and image-guided drug delivery: current concepts and future directions. *Mol Pharm* 2010, 7:1899–1912.

36. Panyam J, Zhou WZ, Prabha S, Sahoo SK, Labhasetwar V: Rapid endo-lysosomal escape of poly(DL-lactide-co-glycolide) nanoparticles: implications for drug and gene delivery. *Faseb J* 2002, 16:1217–1226.

37. Panyam J, Labhasetwar V: Biodegradable nanoparticles for drug and gene delivery to cells and tissue. *Adv Drug Deliv Rev* 2003, 55:329–347.

38. Hughes GA: Nanostructure-mediated drug delivery. *Nanomedicine* 2005, 1:22–30.

39. Bissonnette L, Bergeron MG: Next revolution in the molecular theranostics of infectious diseases: microfabricated systems for personalized medicine. *Expert Rev Mol Diagn* 2006, 6:433–450.

40. Ho YP, Leong KW: Quantum dot-based theranostics. *Nanoscale* 2009, 2:60–68.

41. Lukianova-Hleb EY, Hanna EY, Hafner JH, Lapotko DO: Tunable plasmonic nanobubbles for cell theranostics. *Nanotechnology* 2010, 21:85102.

42. Picard FJ, Bergeron MG: Rapid molecular theranostics in infectious diseases. *Drug Discov Today* 2002, 7:1092–1101.

43. Zou P, Yu Y, Wang YA, Zhong Y, Welton A, Galban C, Wang S, Sun D: Superparamagnetic iron oxide nanotheranostics for targeted cancer cell imaging and pH-dependent intracellular drug release. *Mol Pharm* 2010, 7:1974–1984.

44. Pan D, Carauthers SD, Chen J, Winter PM, SenPan A, Schmieder AH, Wickline SA, Lanza GM: Nanomedicine strategies for molecular targets with MRI and optical imaging. *Future Med Chem* 2010, 2:471–490.

45. Leevy WM, Lambert TN, Johnson JR, Morris J, Smith BD: Quantum dot probes for bacteria distinguish Escherichia coli mutants and permit in vivo imaging. *Chem Commun (Camb)* 2008:2331–2333.

46. Haynes CA, Allegood JC, Park H, Sullards MC: Sphingolipidomics: methods for the comprehensive analysis of sphingolipids. *J Chromatogr B Analyt Technol Biomed Life Sci* 2009, 877:2696–2708.

47. Sanvicens N, Marco MP: Multifunctional nanoparticles–properties and prospects for their use in human medicine. *Trends Biotechnol* 2008, 26:425–433.

48. Ziady AG, Ferkol T, Dawson DV, Perlmutter DH, Davis PB: Chain length of the polylysine in receptor-targeted gene transfer complexes affects duration of reporter gene expression both *in vitro* and *in vivo*. *J Biol Chem* 1999, 274:4908–4916.

49. Verbavatz JM, Ma T, Gobin R, Verkman AS: Absence of orthogonal arrays in kidney, brain and muscle from transgenic knockout mice lacking water channel aquaporin-4. *J Cell Sci* 1997, 110 (Pt 22):2855–2860.

50. Huang X, El-Sayed IH, Qian W, El-Sayed MA: Cancer cell imaging and photothermal therapy in the near-infrared region by using gold nanorods. *J Am Chem Soc* 2006, 128:2115–2120.

51. Alexis F, Basto P, Levy-Nissenbaum E, Radovic-Moreno AF, Zhang L, Pridgen E, Wang AZ, Marein SL, Westerhof K, Molnar LK, *et al.*: HER-2-targeted nanoparticle-affibody bioconjugates for cancer therapy. *ChemMedChem* 2008, 3:1839–1843.

52. Bagalkot V, Farokhzad OC, Langer R, Jon S: An aptamer-doxorubicin physical conjugate as a novel targeted drug-delivery platform. *Angew Chem Int Ed Engl* 2006, 45:8149–8152.

53. Karman J, Tedstone JL, Gumlaw NK, Zhu Y, Yew N, Siegel C, Guo S, Siwkowski A, Ruzek M, Jiang C, *et al.*: Reducing glycosphingolipid biosynthesis in airway cells partially ameliorates disease manifestations in a mouse model of asthma. *Int Immunol* 2010, 22:593–603.

54. Dhar S, Gu FX, Langer R, Farokhzad OC, Lippard SJ: Targeted delivery of cisplatin to prostate cancer cells by aptamer functionalized Pt(IV)

prodrug-PLGA-PEG nanoparticles. *Proc Natl Acad Sci USA* 2008, 105:17356–17361.

55. Farokhzad OC, Cheng J, Teply BA, Sherifi I, Jon S, Kantoff PW, Richie JP, Langer R: Targeted nanoparticle-aptamer bioconjugates for cancer chemotherapy in vivo. *Proc Natl Acad Sci USA* 2006, 103:6315–6320.
56. Zhang Y, Li X, Becker KA, Gulbins E: Ceramide-enriched membrane domains–structure and function. *Biochim Biophys Acta* 2009, 1788:178–183.
57. Gu F, Zhang L, Teply BA, Mann N, Wang A, Radovic-Moreno AF, Langer R, Farokhzad OC: Precise engineering of targeted nanoparticles by using self-assembled biointegrated block copolymers. *Proc Natl Acad Sci USA* 2008, 105:2586–2591.
58. Wang AZ, Gu F, Zhang L, Chan JM, Radovic-Moreno A, Shaikh MR, Farokhzad OC: Biofunctionalized targeted nanoparticles for therapeutic applications. *Expert Opin Biol Ther* 2008, 8:1063–1070.
59. Weng KC, Noble CO, Papahadjopoulos-Sternberg B, Chen FF, Drummond DC, Kirpotin DB, Wang D, Hom YK, Hann B, Park JW: Targeted tumor cell internalization and imaging of multifunctional quantum dot-conjugated immunoliposomes in vitro and in vivo. *Nano Lett* 2008, 8:2851–2857.
60. Leevy WM, Gammon ST, Jiang H, Johnson JR, Maxwell DJ, Jackson EN, Marquez M, Piwnica-Worms D, Smith BD: Optical imaging of bacterial infection in living mice using a fluorescent near-infrared molecular probe. *J Am Chem Soc* 2006, 128:16476–16477.
61. Ziady AG, Kotlarchyk M, Bryant L, McShane M, Lee Z: Bioluminescent imaging of reporter gene expression in the lungs of wildtype and model mice following the administration of PEG-stabilized DNA nanoparticles. *Microsc Res Tech* 2010, 73:918–928.
62. Wu TC, McCarthy VP, Gill VJ: Isolation rate and toxigenic potential of Clostridium difficile isolates from patients with cystic fibrosis. *J Infect Dis* 1983, 148:176.
63. Discher DE, Ahmed F: Polymersomes. *Annu Rev Biomed Eng* 2006, 8:323–341.
64. Discher DE, Eisenberg A: Polymer vesicles. *Science* 2002, 297: 967–973.
65. Discher BM, Won YY, Ege DS, Lee JC, Bates FS, Discher DE, Hammer DA: Polymersomes: tough vesicles made from diblock copolymers. *Science* 1999, 284:1143–1146.

Chapter 2

Multifunctional Tumor-Targeted Nanoparticles for Lung Cancer

Shinji Kuroda,[a] Tomohisa Yokoyama,[a] Justina O. Tam,[b] Ailing W. Scott,[a] Li Leo Ma,[c] Manish Shanker,[a] Jiankang Jin,[a] Corbin Goerlich,[a] David Willcutts,[a] Jack A. Roth,[a] Konstantin Sokolov,[b,d] Keith P. Johnston,[c] and Rajagopal Ramesh[a,e,f,*]

Departments of [a]Thoracic and Cardiovascular Surgery, and [d]Imaging Physics, University of Texas, MD Anderson Cancer Center, 1515 Holcombe Blvd, Houston, Texas 77030, USA
Departments of [b]Biomedical Engineering, and [c]Chemical Engineering, University of Texas, Austin, Texas 78712, USA
Department of [e]Pathology and [f]Graduate Program in Biomedical Sciences, University of Oklahoma Health Sciences Center, Oklahoma City, Oklahoma 73104, USA
[*]rajagopal-ramesh@ouhsc.edu

Lung cancer is a major health problem in the United States. Despite the development of molecular-targeting agents such as epidermal growth factor receptor (EGFR) tyrosine kinase inhibitors and advances made in conventional chemoradiotherapy and surgical therapy, the overall five-year survival rate of lung cancer patients remains poor and is less than 15%. Recently, in an attempt to improve diagnosis and therapy, novel technologies such as nanotechnology have emerged for application in medicine. One major goal of nanomedicine is to develop multifunctional nanoparticles that can be applied for diagnosis and imaging of cancer as well as therapy for cancer. As a result, a number of nanoparticle agents of various

Pulmonary Nanomedicine: Diagnostics, Imaging, and Therapeutics
Edited by Neeraj Vij

ISBN 978-981-4316-48-4 (Hardback), 978-981-4364-14-0 (eBook)
www.panstanford.com

compositions, sizes, and shapes are being developed. A majority of the nanoparticles, however, are in preclinical studies, and only a few of them have advanced to early clinical testing. Efforts in several laboratories, including our own laboratory, are in developing tumor-targeted multifunctional metal-iron-oxide-based nanoparticles for imaging and therapy of lung cancer. In this article, we will discuss various nanomaterial-based nanoparticles, including our own tumor-targeted multifunctional nanoparticles that are being developed for lung cancer. Readers are encouraged to review additional literature to obtain more information on nanomaterials and their application in nanomedicine and cancer therapy.

2.1 Introduction

Lung cancer is the most common cause of death across genders, races, and ethnicities [1, 2]. Approximately 85% of lung cancers are non-small-cell lung cancer (NSCLC), 75% of which are metastatic at diagnosis. This is because most symptoms do not appear until the disease is advanced. The treatment of NSCLC includes conventional cytotoxic chemotherapy and radiotherapy. Advances made in the treatment strategy, such as the emergence of molecular targeting agents, have demonstrated improvements in lung cancer patient survival. However, the overall five-year survival of patients diagnosed with NSCLC remains less than 15% [2]. One major challenge that remains with many therapeutic regimens is the inability to deliver cancer drugs specifically to tumors while achieving high intratumoral drug concentrations. As a result, systemic administration of anticancer drugs results in poor drug accumulation within tumors, contributing to treatment failure [3]. Thus, it is important to develop novel therapies that are tumor targeted, effective in controlling tumor growth while maximizing tumor uptake.

Advancements in the understanding of the molecular cause of lung cancer have resulted in the identification of the EGFR as an important player in lung cancer. EGFR is a member of the erythroblastic leukemia viral oncogene homolog (ErbB) family of receptor tyrosine kinases and approximately 50–80% of NSCLC demonstrate overexpression of the EGFR pathway [4–6]. This receptor has been shown to activate a variety of cellular signaling pathways and play an important role in cell proliferation, invasion, metastasis, and angiogenesis [7]. Thus, the elimination or reduction of EGFR expression in lung cancer cells using EGFR- targeted drugs can lead to cancer cell death. Based on this concept, several EGFR-targeted biological (cetuximab) and synthetic small-molecule (erlotinib and gefitinib) inhibitors have been developed and are currently undergoing testing for lung cancer. Cetuximab (Erbitux; C225) is a humanized monoclonal antibody that targets the extracellular domain of EGFR while erlotinib (Tarceva) and gefitinib (Iressa) target the intracellular kinase domain of EGFR [8–11].

Studies have shown that lung cancer patients harboring mutations in EGFR responded to EGFR therapy, showed clinical response, and had improved survival. These results suggest that this subpopulation of lung cancer patients should receive EGFR-targeted therapy. However, recent studies have reported lung cancer patients without EGFR mutations also respond to treatment with EGFR inhibitors. These studies indicate that the expression or mutational status of EGFR alone is not a reliable marker for determining the outcome of EGFR inhibitor–based therapy; therefore, this therapy should be administrated to all patients with lung cancer, regardless of the patient's EGFR mutational status. However, a potential drawback of EGFR-based therapy, as with other therapies, is the possibility that patients will develop resistance to treatment. For example, some patients who initially respond to gefitinib or erlotinib therapy develop resistance after prolonged treatment. This problem is further perpetuated since current treatment techniques do not allow for a rapid and noninvasive determination of whether patients receiving EGFR-based therapy are responding to treatment and when it is appropriate to discontinue therapy or change therapy. As a result, patients with lung cancer may receive an ineffective treatment for extended periods of time. Thus, it is of importance to develop tumor-targeted therapies that not only maximize effectiveness and tumor uptake but also allow for noninvasive therapeutic evaluation.

To overcome these limitations of cancer-targeted therapies, novel technologies such as nanotechnology are being developed and tested in the laboratory. Nanotechnology as defined by the National Cancer Institute (NCI; www.nci.gov) is the field of research that deals with the engineering and creation of materials that are less than 100 nm in size, especially single atoms or molecules. Nanotechnology for cancer is being developed to facilitate rapid monitoring of drug delivery and assess the therapeutic effect of the drug noninvasively. Synthesis of nanoparticles of various compositions, sizes, and shapes is an important subfield of nanotechnology. However, nanoparticles that are around 50 nm in size have been reported to be the most efficiently recognized and internalized by cells, resulting in an optimum therapeutic effect [15]. This, however, also raises concern for nonspecific toxicity, the phenomenon of noncancerous cells being able to internalize these nanoparticles. Therefore, to achieve tumor cell specificity and minimize normal cell toxicity, investigators are utilizing a plethora of scientific information (phenotype, genotype, biomarkers, and biochemical properties) available about cancer cells to engineer tumor-targeted nanoparticles. The most common approach is the coating of the outer surface of the nanoparticles with ligands (peptide, antibodies, DNA, etc.) that are uniquely expressed and recognized by tumor cells or ligands that are overexpressed by tumor cells compared to normal cells. The inner core of the nanoparticles can further be designed to carry drugs (chemotherapy, siRNA, microRNA [miRNA], and DNA) that are selectively released and activated only in tumor cells. The nanoparticles thus can be designed for delivering anticancer drugs specifically to cancer cells, thereby minimizing toxicity to normal cells, to carry molecules that will enable the detection of cancer cells by molecular imaging noninvasively without surgery, and to determine responses to treatment rapidly and noninvasively by imaging [12–14]. Thus, the application of nanotechnology in cancer medicine will prove to be beneficial in the clinic for patients diagnosed with cancer as these aforementioned properties of nanoparticles can address many common cancer therapy problems faced by clinicians.

In this article we will discuss the utilization of biomarkers for developing targeted nanoparticles, describe various nanomaterial-based nanoparticles that are being developed for lung cancer, and finally explain the approach we have undertaken for the development of tumor-targeted multifunctional nanoparticles.

2.2 Biomarkers for Tumor Targeting

Table 2.1 shows some of the biomarkers that have been reported to be overexpressed on tumor cell surfaces, blood vessel endothelium, and extracellular matrices (ECMs) of tumor tissues [16, 17]. These and other biomarkers, as discussed below, have been exploited for the development of tumor-targeted nanoparticles and achieving tumor specificity.

Tumors usually have relatively porous blood vessels and poor lymphatic drainage. Nanoparticles can extravasate into tumor tissues via these highly disorganized and "leaky" blood vessels by the enhanced permeability and retention (EPR) effect [18]. The increased vessel permeability and the dysfunctional lymphatic drainage in tumor tissues allow nanoparticles to accumulate in tumor lesions. Although passive targeting like EPR effect is actually used clinically, it is also true that there are some limitations in this targeting strategy. One way to overcome these limitations is to attach ligands to nanoparticles targeted against a specific cancer cell biomarker so that nanoparticles can specifically bind and be internalized into tumor cells. For example, EGFR is overexpressed in a variety of cancers such as lung, breast, and head and neck. Her2/neu, which is another member of the ErbB receptor family, is also overexpressed in breast and ovarian cancers. In addition, folate receptors and transferrin receptors are overexpressed in several types of cancers. Nanoparticles coated with antibodies or other ligands targeting these receptor molecules have demonstrated improved drug delivery, increased drug accumulation in tumors, and increased therapeutic efficacy with reduced toxicity to normal tissues [19–22].

Table 2.1 Biomarkers overexpressed in tumor tissues

Tumor cell surface	Tumor vessel endothelium	Tumor ECM
EGFR	Integrins($\alpha_v\beta_3$ and $\alpha_v\beta_5$)	Hyaluronan
Her2/neu	Tumor endothelial markers	Heparan sulfate
Folate receptors	Fibronectin ED-B	Chondroitin sulfate
Transferrin receptors		

Another approach for successful tumor targeting is to facilitate circulating nanoparticles' accessibility from tumor blood vessels to the ECM of tumor cells and within tumor cells themselves. This has been achieved by targeting molecules that are overexpressed in the endothelium of tumor vessels. For example, the $\alpha_v\beta_3$ and $\alpha_v\beta_5$ integrins are highly expressed in the endothelium of tumor vessels; and tumor-homing peptides targeting these integrins, such as arginine-glycine-aspartic acid (RGD), asparagines-glycine-arginine

(NGR), and F3, have been identified and tested for targeted drug delivery [23, 24]. Curnis *et al.* showed coupling of RGD peptide to tumor necrosis factor (TNF) alpha, which resulted in *in vivo* anticancer activity, which was enhanced when combined with melphalan [25]. Furthermore, the effective doses of RGD-TNF to produce a therapeutic effect were 1,000 times lower than the effective doses of TNF alone. Paoloni *et al.* demonstrated adeno-associated virus (AAV)-phage vector that delivered RGD-TNF produced clinical responses in a canine tumor model [26]. These studies demonstrated improved therapeutic efficacy resulting from tumor-targeted drug delivery.

In addition to tumor cell surface markers and tumor endothelium markers, some ECM such as hyaluronan, heparan sulfate, and chondroitin sulfate are also overexpressed in tumor tissues and have been studied for targeted drug delivery. Nanoliposomes containing doxorubicin or mitomycin C coated on the outside with hyaluronan demonstrated increased half-life and circulation time *in vivo*, enhanced tumor targeting to hyaluronan receptor expressing tumors, and greater antitumor activity compared to the antitumor activity produced by nanoliposomes without hyaluronan [27, 28].

2.3 Nanotechnology in Medicine

2.3.1 Development of Nanoparticles for Lung Cancer and Other Medical Applications

Table 2.2 shows nanoparticle agents that have been tested in clinical trials or preclinical trials for lung cancer, and Table 2.3 shows nanoparticle agents that have been already marketed or tested in clinical trials for other medical applications.

Table 2.2 Nanoparticle agents for lung cancer

Study phase	Product (description)	Supplementary note
Phase III	Abraxane® (ABI-007) (Paclitaxel albumin nanoparticles)	ABI-007 + Carboplatin vs. Taxol + Carboplatin
Phase II	L9NC (Aerosolized liposomal camptothecin)	Inhalation drug

Study phase	Product (description)	Supplementary note
Phase I	Fus1 conjugated DOTAP:chol. (Liposomal therapeutic gene)	Fus1: tumor suppressor gene
Preclinical	MSCs with iron oxide nanoparticles [29]	MRI contrast agent
Preclinical	EGF-modified gelatin nanoparticles with FITC [19] or cisplatin [30]	Inhalation drug

Abbreviations: DOTAP:chol., 1,2-dioleoyl-3-trimethylammonium-propane:cholesterol; EGF, epidermal growth factor; FITC, fluorescein iso-thiocyanate; MRI, magnetic resonance imaging; MSCs, mesenchymal stem cells.

Table 2.3 Nanoparticle agents for other medical applications

Study phase	Product (description)	Use
Marketed	Dosil® (Liposomal doxorubicin)	Ovarian cancer
Marketed	Abraxane® (ABI-007) (Paclitaxel albumin nanoparticles)	Metastatic breast cancer
Marketed	AmBisome® (Liposomal amphotericin B)	Fungal infection, etc.
Marketed	DounoXome® (Liposomal dounorubicin)	Kaposi's sarcoma
Phase III	Myocet® (Liposomal doxorubicin)	Metastatic breast cancer
Phase II	Feraheme® (Ferumoxytol) (Iron oxide nanoparticles)	MRI contrast agent
Phase I	Aurimune® (CYT-6091) (TNFα-PEG-colloidal gold nanoparticles)	Solid tumors
Phase I	CALAA-01 (siRNA-cyclodextrin polymer nanoparticles)	Solid tumors
Phase I	Nanoxel (Paclitaxel nanoparticles)	Breast cancer

The development of drug delivery systems based on nanotechnology is being tested for diseases like cancer, diabetes, and infections and in gene therapy. The main advantages of this modality are specific targeting of the drug to the lesion or source of the disease and enhanced safety profile. Nanotechnology has also been used in diagnostic medicine to develop contrast agents, fluorescent dyes, and magnetic nanoparticles [31].

Abraxane® (ABI-007) — albumin-bound paclitaxel — was initially tested for the treatment of metastatic breast cancer. ABI-007 demonstrated high tumor penetration and strong antitumor activity in human breast tumor xenografts established in athymic nude mice. Preclinical results lead to the testing of ABI-007 in phase I and II clinical trials that was followed by a large international randomized phase III clinical trial for metastatic breast cancer. Results from the phase III trial demonstrated that ABI-007 treatment was superior to standard paclitaxel treatment for both the overall response rate (ORR; 33% vs. 19% respectively; $p = 0.001$) and time to tumor progression (TTP; $p = 0.006$) [32]. Furthermore, treatment-related toxicity was significantly lower in patients receiving ABI-007 than in patients receiving paclitaxel. ABI-007 treatment is currently used in breast cancer patients diagnosed with metastatic disease and who have failed combination chemotherapy and/or in patients when the disease relapses within six months of adjuvant chemotherapy.

The findings from the breast cancer studies have resulted in the testing of ABI-007 for lung cancer. The combination of paclitaxel and carbolatin is a standard treatment for advanced NSCLC. A phase II trial of ABI-007, carboplatin, and bevacizumab (monoclonal antibody for vascular endothelial growth factor A [VEGF-A] for advanced NSCLC (stage IIIB/IV) demonstrated that the combination of these drugs produced encouraging survival results (response rate: 31% [with stable disease rate of 54%]; median progression-free survival: 9.8 months; median survival: 16.8 months) and well tolerable and manageable adverse event (grades 3 and 4: neutropenia [54%], fatigue [17%]) [33]. Based on the results, a randomized phase III trial of ABI-007 plus carboplatin compared with taxol plus carboplatin as the first-line therapy in patients with advanced NSCLC is ongoing, and the results are yet to be reported.

Preclinical studies from our laboratory using a cationic lipid DOTAP:chol.-based nanoparticles demonstrated that these nanoparticles carrying therapeutic genes such as p53, Fus1, and melanoma differentiation–associated (mda) gene-7/interelukin (IL)-24 effectively delivered the therapeutic genes to metastatic tumors and produced therapeutic effects when administered systemically into mice [34–38]. Furthermore, no significant treatment-related toxicity was observed. Based on these preclinical results, we have recently initiated a phase I clinical trial for the systemic treatment of stage IV NSCLC patients who have failed chemotherapy [39]. In this clinical trial, NSCLC patients are systemically treated with DOTAP: chol. nanoparticles carrying the Fus1 tumor suppressor gene. The objective of this trial is to determine the maximum-tolerated dose (MTD). Initial results show DOTAP:chol.-Fus1 nanoparticle treatment is well tolerated with no observable treatment-related toxicity. The final results from this trial will be known when completed.

In addition to ABI-007 and DOTAP:chol. nanoparticles, several other nanoparticle agents for medical applications are already in clinical use. Additionally, a number of new nanoparticles are being developed for therapy and diagnosis for diverse human diseases.

2.3.2 Classes of Nanoparticles

Liposomes, which were discovered in the 1960s, were the original types of nanoparticles that have been explored as effective drug delivery systems. Chemotherapeutic drugs or other toxic drugs such as amphotericin, when used as liposomal drugs, produce much better efficacy and safety compared to conventional preparations. Several liposomal drugs have been already used in clinical practice (Table 2.3). However, one major limitation of liposomes is that they are rapidly degraded and quickly cleared from the blood stream by the liver, thus reducing the availability of the drug. As a result, the accumulation of the drug at concentrations at the diseased site is very low, resulting in lack of therapeutic activity. Progress has been made to overcome this challenging issue by modifying the liposome-based nanoparticles, and liposomes still remain promising tools for drug delivery [40].

Nanopores, which were designed in 1997, consist of wafers with a high density of pores, 20 nm in diameter [41]. The pores allow oxygen, glucose, and other products like insulin to pass through while preventing immunoglobulin and cells from doing the same. Because of this, nanopores are used as devices to protect transplanted tissues from the host's immune system. Also, nanopores can be used for DNA sequencing since these particles exhibit the ability to differentiate DNA strands based on differences in base pair sequences [42].

Fullerenes, which were discovered in 1985, consist entirely of carbon [43]. The most common form of fullerenes, "Buckyball," is shaped like a soccer ball with 20 hexagons and 12 pentagons. Fullerenes have been investigated for drug delivery systems of anticancer agents, antibiotics, and antiviral drugs and can also be used in cancer therapy utilizing fullerenes' ability to induce reactive oxygen species (ROS) activity during photosensitization [44–46].

Carbon nanotubes, which were discovered in 1991, are tubular carbon structures [47]. Nanotubes can be single-walled carbon nanotubes or multiwalled carbon nanotubes and vary in the length, ranging from 1 to a few micrometers. They are characteristically strong and stable, which allows nanotubes to be used as carriers for drugs like amphotericin B [48] transcription interference molecules such as siRNA [49].

Quantum dots are nano-sized crystals and have played an important role as imaging tools in medicine. The most notable characteristic of quantum dots is their brightness. It is estimated that quantum dots are 20 times brighter and 100 times more stable than traditional fluorescent reporters [50]. Quantum dots are currently being used as a therapeutic and diagnostic tool. These particles, which are functionalized with polyethylene glycol (PEG) and an antibody that targets specific molecules, have demonstrated their accumulation and retention in xenografts of mouse models [51]. Quantum dots can also be used for sentinel node imaging because of their strong near-infrared (NIR) intensity. Sentinel node imaging has been adopted for several malignant tumors, including melanoma, breast, lung, and gastrointestinal cancers and aids in planning of therapy [52].

Nanobubbles (or microbubbles) are used to deliver therapeutic drugs. They consist of a desired therapeutic drug that is incorporated into nanobubbles designed to target tumor tissues. These nanoparticles are administered in the blood stream and reach tumor lesions by utilizing extravasations from vessels surrounding the tumor and accumulate at the site of tumor. This is followed by microbubble formation, which can be visualized with ultrasound. When the site is located and then targeted with high-intensity focused ultrasound, the high-frequency sound waves lyse the bubbles, resulting in the release of the drug. This system enables highly specific tumor-targeting drug delivery, contributes to increased efficacy, and reduces toxicity. Moreover, these materials can be used as contrast agents for ultrasound [56, 57].

Paramagnetic nanoparticles can be used for both therapeutic and diagnostic purposes. Paramagnetic iron oxide nanoparticles, which are used as contrast agents in MRI, produce greater contrast than conventional contrast agents such as gadolinium. Paramagnetic nanoparticles conjugated with antibodies to HER-2/neu, which are expressed on breast cancer cells, have been used with MRI to detect breast cancer cells [58]. Other paramagnetic nanoparticles conjugated with luteinizing hormone releasing hormone (LHRH) have been demonstrated to detect breast cancer cells *in vivo* [59]. Thus, targeting of these nanoparticles via antibodies enables the identification of specific tumors, organs, and tissues [60, 61]. Magnetic nanoparticles are also used as therapeutic agents. For example, under optimal conditions, iron oxide nanoparticles, which accumulate in the targeted site, can generate heat to the level of inducing cell death [62].

Plasmonic gold nanoparticles are also an attractive type of nanoparticle for cancer diagnosis and therapy because they are relatively easy to synthesize and gold has already been approved and used for the treatment of rheumatoid arthritis [63]. Nanoshells are a type of spherical nanoparticle that consists of a dielectric core such as silica covered by a thin metallic shell such as gold and are usually used in thermal ablation. Nanoshells can be targeted to desired tissues by immunological means and are currently used as a novel modality of targeted therapy [53]. This subclass of nanoparticles are also engineered to absorb light near the infrared region and hardly become absorbed in normal tissue. Because of this characteristic, nanoshells are optimal for deep tissue treatment. Illumination of nanoshells is usually done by laser light. This light is absorbed by nanoshells and converted into heat, which increases the temperature of the nanoshells to over 30°C. This nanoshell-based photothermal ablation therapy demonstrated an over 90% success rate in causing tumor remission in mice [54, 55]. Other types of NIR-absorbing plasmonic nanoparticles such as nanorods and nanocages have shown high efficacy in photothermal killing of cancer cells in experiments *in vitro*. It was also demonstrated that EGF receptor–mediated aggregation of spherical gold nanoparticles can be used for molecular-specific photothermal therapy with NIR laser.

Gold nanoparticles have been used as optical contrast agents because of their ability to strongly scatter and absorb visible and NIR light [64]. They have also been tested for use as therapeutic agents. For example, a phase I clinical trial of colloidal gold nanoparticles conjugated with tumor necrosis factor (TNF) (CYT-6091) was conducted on the basis of preclinical studies (Table 2.3) [65, 66]. In addition, it has been demonstrated that gold nanoparticles, when combined with irradiation, can produce a synergistic effect in mice [67].

2.3.3 Delivery Methods of Nanoparticles to Targeted Regions

2.3.3.1 Systemic Administration

One of the most significant issues for the use of nanoparticles as therapeutic or diagnostic tools in the body is how to target and accumulate the necessary amount of nanoparticles in the tumor *in situ* or in tissues like lymph nodes after systemic administration. Solving this issue requires four components [68]. First, nanoparticles must be able to navigate the bloodstream. Nonspherical particles are more likely to attach to the vascular wall than spherical particles. Second, nanoparticles must be designed to evade biological barriers. For example, nanoparticles of the right size go through the cancer-associated capillary wall and localize preferentially in cancer lesions. Third, nanoparticles must also be designed to localize in a site- and cell-specific manner. Nanoparticles of different sizes are taken up into cancer cells with different efficiency. Finally, nanoparticles need to target the biological pathways for treatment. For instance, some nanoparticles carry drugs to cancer cells, while other nanoparticles are themselves potential anticancer agents.

The size, shape, and surface chemistry of nanoparticles can influence their circulation period and distribution pattern in the body, but the final distribution of particles is reported to be 80–90% in the liver, 5–8% in the spleen and 1–2% in the bone marrow [69]. Large (>100 nm) particles are easily captured by the liver and spleen, and, even though they move into the ECM from blood vessels, their accumulation is usually restricted to the perivascular space, with little permeation into cells. On the other hand, small (<10 nm) particles are readily internalized into cells but are also rapidly removed through renal clearance. Therefore, 10–100 nm (especially around 50 nm) sized particles are considered to be optimal for systemic administration [15, 70, 71]. Surface chemistry indicates the efficacy and mechanism of nanoparticles' uptake by cells and their total biodistribution in the body [72].

Within minutes of their systemic administration, nanoparticles are challenged by the macrophages of the reticulo-endotherial system (RES) [69]. Depending on their surface hydrophobicity and charge, nanoparticles may be opsonized by plasma proteins such as albumin, apolipoprotein, immunoglobulins, and fibrinogen, which promotes their recognition and clearance by cells of the RES [73]. Many efforts have been made to engineer nanoparticles of size and surface chemistry that minimize their opsonization and macrophage-mediated clearance, thus increasing their circulation time. One such effort is the surface coating of nanoparticles with amphiphilic polymeric surfactants, such as PEG, which significantly reduces the nanoparticles' interactions with plasma proteins and minimizes their internalization and clearance by macrophages [74].

Macrophages are a critical element of the body's immune system, and introduction of nanoparticles into the body may affect macrophage function. For example, it is known that nanoparticles can activate phagocytotic and cytokine-releasing functions of macrophages [75].

2.3.3.2 Local Administration via Inhalation

Direct drug administration to the lung via inhalation offers several theoretical advantages over systemic delivery, including the possibility of regional drug delivery to the lung and airways with lower doses and less systemic toxicity, avoiding first-pass metabolism of the drug in the liver and the use of a noninvasive "needle-free" delivery system [76]. The local delivery of therapeutic drugs via inhalation for primary or metastatic lung cancer could increase drug exposure of the lung cancer and minimize systemic toxicity.

One of the most important factors of aerosol dose and distribution in the lung is the particle size [76]. Particles of 5–10 nm are mainly deposited in the oropharynx and large airways, and particles of 1–5 nm are deposited in the small airways and alveoli with more than 50% of particles of 3 nm being deposited in the alveoli. The different sizes of aerosol particles may be used for lung cancer therapy based on the location of the cancer, whether near the central airways or in the alveolar regions.

The clinical trials of aerosolized chemotherapy started with 5-fluorouracil (5-FU) in patients with NSCLC [77], followed by aerosolized chemotherapy with camptothecin, doxorubicin, and cisplatin [78–80]. Among these, the aerosolized chemotherapeutic drug closest to clinical is liposomal 9-nitro-20 (S)-camptothecin (L9NC) (Table 2.2) [78]. Based on the result of a phase I clinical trial that showed aerosolized L9NC delivered via inhalation has the potential for a therapeutic effect against primary lung cancer with tolerable toxicity, the phase II trial of aerosolized L9NC in patients with NSCLC has been conducted (www.ClinicalTrials.gov). Further studies on the efficacy, safety, and pharmacokinetics are required to determine whether the aerosolized administration of chemotherapeutic drugs is effective for lung cancer treatment. Still, the ability to allow direct exposure of therapeutic drugs to cancer cells and reduction of systemic toxicity could make aerosolized chemotherapy via inhalation an attractive method against selected types or locations of lung cancers [81].

2.3.3.3 Toxicity of Nanoparticles

Progress in engineering nanostructures, elucidation of their unique properties, and broadening of their applications has made nanotechnology an interesting research area. A wide variety of nanoparticles have been developed as therapeutic or diagnostic tools, and their efficacy has been demonstrated. Perhaps the most fundamental characteristic of nanoparticles is their small size (<100 nm), which allows them to reach their targeted areas and deliver drugs into targeted cells more easily. However, this also means that, without strict tumor targeting, nanoparticles can enter normal cells and cause toxicity.

One mechanism for nanoparticle-mediated toxicity reported is the induction of oxidative stress, which can induce DNA damage by increasing intracellular ROS [82]. Activation of cellular signaling pathways resulting in the release of pro-inflammatory cytokines has also been shown to contribute to nanoparticle-mediated toxicity [83]. The production of acute inflammatory response in response to nanoparticles treatment can result in DNA damage via epigenetic alterations [84] and plays a role in promoting carcinogenesis [85]. In addition to their composition, the high surface area of nanoparticles can promote the generation of ROS.

Gold nanoparticles are capable of inducing DNA damage indirectly through oxidative stress and have been shown to be dependent on the size of the particles and/or cell type. For example, gold nanoparticles that are 3–8 nm in size have been shown to reduce ROS in a time- and dose-dependent manner and be noncytotoxic and nonimmunogenic [86]. In contrast, a recent study demonstrated that gold nanoparticles that are 20 nm in size induce statistically significant oxidative DNA damage at concentrations as low as 25 μg/ml [87].

Iron has been associated with cancer for a long period of time, and several pathways for iron-induced carcinogenesis have been demonstrated based on oxidative stress–causing lipid peroxidation and direct damage to DNA and proteins [88, 89]. Spindle cell sarcoma and pleomorphic sarcoma in rats have been associated with iron overload following intramuscular injections of iron-dextran complex [90]. Therefore, overloading iron oxide nanoparticles in cells may have negative effects on cells, such as carcinogenesis. However, the effects of iron-based nanoparticles on cell toxicity are yet to be determined.

The majority of studies till date have focused on investigating nanoparticle-mediated toxicity using *in vitro* models. However, a lack of correlation between *in vitro* and *in vivo* effects has recently been demonstrated [91]. Thus, more detailed *in vivo* studies investigating the effects of nanoparticles are required. The outcome of such detailed studies will lead to a better understanding of nanoparticles' behavior *in vivo* and provide a basis for predicting toxicity [92].

2.4 EGFR-Targeted Hybrid Plasmonic Magnetic Multifunctional Nanoparticles

2.4.1 Structure of Nanoparticles

Plasmonic nanoparticles such as gold and silver can be used for combined imaging and photothermal therapy of cancer cells. Furthermore, it has been demonstrated that plasmonic nanoparticles can be combined with other inorganic materials, such as iron oxide for MRI, to form hybrid nanomaterials that provide easily detectable signals in more than one imaging modality [93].

The selection of biomarkers for conjugation with nanoparticles for proper tumor targeting is one of the most crucial factors for creating new nanoparticle agents. Biomarkers vary for each cancer type, and good biomarkers show significant overexpression in their respective cancer type. For example, EGFR is overexpressed in 50–80% of NSCLC, making it a desirable biomarker candidate for NSCLC.

We have developed novel, tumor-targeted multifunctional nanoparticles, which are 50–80 nm in size and consist of a inner paramagnetic iron core that is surrounded by a gold layer and are functionalized with a monoclonal anti-EGFR antibody (cetuximab) (Fig. 2.1) [64, 93, 94]. The iron core will facilitate imaging via MRI. The outer gold shell serves as a platform for attaching several ligands on the surface of the nanoparticles as well as is useful for optical imaging. The anti-EGFR antibody will both serve as a targeting ligand for EGFR overexpressing cancer cells and function as a therapeutic by inhibiting EGF-mediated cell signaling. Thus, our nanoparticles are multifunctional in their properties that need to be tested *in vitro* and *in vivo* against lung cancer models that overexpress EGFR.

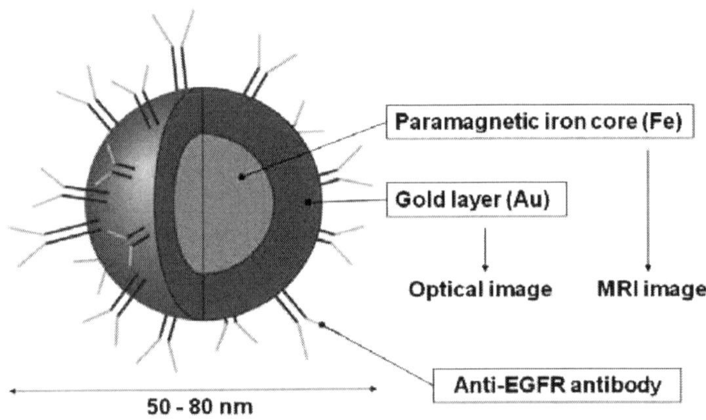

Figure 2.1 The structure of EGFR-targeted hybrid plasmonic magnetic nanoparticles. See also Color Insert.

2.4.2 Therapeutic Function of EGFR-Targeted Nanoparticles

2.4.2.1 Inhibition of EGFR signaling pathway

EGFR activates several signaling pathways and contributes to cell proliferation, invasion, inhibition of apoptosis, metastasis, and angiogenesis. Anti-EGFR monoclonal antibody (cetuximab) binds to the extracellular ligand-binding domain with an affinity five times greater than that of natural ligands like TGF-α and EGF [95]. Binding of cetuximab prevents dimerization and subsequent activation by autophosphorylation of the receptor in the intracellular kinase domain. One of the intriguing things about cetuximab's efficacy against NSCLC is that cetuximab is effective not only on EGFR mutant NSCLC lines known for sensitivity to EGFR-tyrosine kinase inhibitors, such as gefitinib and erlotinib (L858R and delL747-T753insS) but also on those cells that harbor T790M mutation and are resistant to tyrosine kinase inhibitors [96]. At the moment, cetuximab is being used just against patients of colorectal cancer and head and neck cancer and is not indicated for use on patients with lung cancer. However, more recently, testing of cetuximab against lung cancer either as monotherapy or in combination with conventional chemoradiotherapy or tyrosine kinase inhibitors has been completed [6, 97–99].

In the same way as cetuximab, EGFR-targeted nanoparticles [100], after binding to EGFR, blocked EGFR activity and downstream signaling such as AKT, p38MAPK, and p44/42MAPK. We have verified that the effect was observed at 15 minutes after treatment and lasted over 60 minutes in not only EGFR-mutant NSCLC lines but also those NSCLC lines with wild-type EGFR that tyrosine kinase inhibitors are not candidates for [101]. On the other hand, in the normal cells and EGFR-null cells of the lung, this effect was not observed, which means inhibition of EGFR-signaling pathway by EGFR-targeted nanoparticles depends heavily on the expression level of EGFR.

Moreover, after binding to EGFR, nanoparticles are taken up into cells along with EGFR by receptor-mediated endocytosis, which leads to decreased EGFR availability [102]. It has been verified that 40–50 nm nanoparticles have the greatest influence on the signaling process essential for basic cell functions, including cell death, although nanoparticles within 2–100 nm size range were found to alter that signaling process [15]. This means nanoparticles of well-defined sizes can selectively induce membrane receptor internalization to down-regulate their expression level and alter the downstream signaling responses [15] (Fig. 2.2).

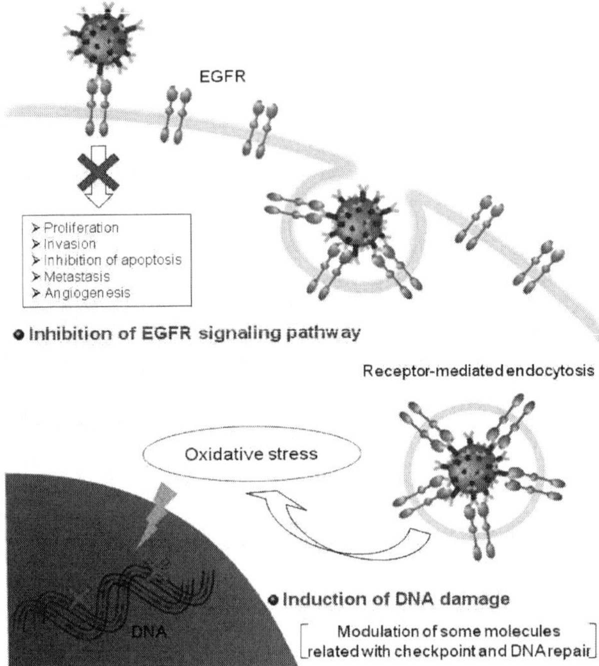

Figure 2.2 The mechanism for therapeutic effects of EGFR-targeted nanoparticles. See also Color Insert.

2.4.2.2 Induction of DNA damage

In medicine, nanoparticles were developed to carry drugs to cancer cells, but it has been proven recently that several types of nanoparticles, magnetic nanoparticles for one, have their own cytotoxic activity. An important mechanism of nanoparticles' inherent cytotoxic activity is oxidative stress resulting in the production of ROS, an internal metabolic event that can induce single-stranded or double-stranded DNA breaks.

ROS are produced as a normal product of cellular metabolism of oxygen. A major source of ROS is hydrogen peroxide (H_2O_2), which is converted from superoxides produced by mitchondoria. In addition to the endogenous source, several exogenous sources such as UV rays, heat exposure, and especially ionizing radiation can produce excessive ROS. Cells normally have the ability to defend themselves against ROS through antioxidants such as ascorbic acid (vitamin C), tocopherol (vitamin E), uric acid, and glutathione. However, excessive ROS, which induces excessive DNA damage, leads to cell death through apoptosis [103–105].

The most widely accepted model of a DNA damage-induced signaling pathway is the ionizing radiation model. DNA damage induced by ionizing radiation immediately activates transducers such as ataxia-teleangiectasia mutated (ATM) and ataxia-teleangiectasia receptor (ATR) [106]. This leads to downstream signaling, playing an important role in the DNA damage checkpoint by controlling several key proteins such as p53, Mdm2, BRCA1, Chk2, MDC1, NBS1, and H2AX [106]. In recent studies, nanoparticles that consisted of carbon or metal were shown to induce DNA damage responses by modulating molecules related to checkpoint and DNA repair in a way similar to radiation [105].

One key effecter molecule activated after DNA damage is p53, a tumor suppressor gene. p53 is responsible for arresting the cell cycle while repairing damaged DNA, and if the damage present is extensive, then p53 triggers apoptosis in order to eliminate the damaged cells. In a recent study, gold nanoparticles have been shown to down-regulate a number of DNA repair genes, including BRCA1, a tumor suppressor gene related to hereditary breast cancer [87]. BRCA1 is also a crucial molecule for the DNA damage repair process, which plays just as important a role in the cell cycle checkpoint and transcription regulation as it plays in DNA repair [107].

We have shown that hybrid plasmonic magnetic nanoparticles conjugated with cetuximab induce DNA damage, inhibit BRCA1 expression, and modulate some molecules related to the cell cycle checkpoint or DNA repair in lung cancer cells. This means EGFR-targeted nanoparticles can induce and lead to the accumulation of damage in cells by inhibiting proper DNA repair systems.

2.4.3 Diagnostic Function of EGFR-Targeted Nanoparticles

Gold nanoparticles exhibit intense colors in the visible and NIR region and are novel cellular-imaging tools *in vitro* and *in vivo*. Gold nanoparticles conjugated with anti-EGFR antibody enable us to optically detect EGFR-overexpressing cancer cells or precancerous cells. This means we can not only diagnose cancer or precancerous cells overexpressed EGFR from a patient's biopsy samples but also know quickly if those cells detected with gold nanoparticles conjugated with anti-EGFR antibody may be sensitive to EGFR-targeting therapeutic drugs [64, 108].

The paramagnetic iron core enables us to diagnose EGFR-overexpressed lung cancers, monitor the change of the tumor size during and after treatment of chemotherapy or radiotherapy, and detect the nanoparticles' distribution around cancers and throughout the body as it is a contrast agent for MRI. Preclinical data for this component in EGFR-targeted nanoparticles is not substantial yet, but the utility of iron oxide nanoparticles in medicine as contrast agents of MRI has been investigated in clinical trials (Table 2.2) [109].

2.5 Conclusions

Nanotechnology, although still in its early phase, has significantly advanced the field of medicine in the area of diagnosis and therapy. In the last few years, there has been tremendous interest and rapid growth in the field of nanomedicine all around the world. As a result, a number of new nanoparticles using novel nanomaterials are being developed. Additionally, with the explosion of information on biomarkers, investigators are developing nanoparticles that are functionalized with unique ligands that will selectively target lung tumor cells and produce an enhanced therapeutic response with minimal toxicity to the surrounding normal tissues. Furthermore, some of the nanoparticles similar to our nanoparticles will be multifunctional in their properties, which will be useful for diagnosis, imaging, and therapy. Despite these advances made, it is anticipated that only a few among several of these nanoparticles will pass the preclinical testing with minimal toxicity and will be made available for lung cancer therapy in the clinic.

Acknowledgments

The authors thank the present and past laboratory members who have contributed to this research. This work was supported by grants from Joans Legacy: United Against Lung Cancer, NIH/NIBIB R03 EB009182, and Lung SPORE P50 CA70970.

References

1. Parkin, D.M. (2001) Global cancer statistics in the year 2000. *Lancet Oncol*, **2**, 533–543.
2. Jemal, A., Siegel, R., Ward, E., Hao, Y., Xu, J., Murray, T., and Thun, M.J. (2008) Cancer statistics, 2008. *CA Cancer J Clin*, **58**, 71–96.
3. Sharma, S., White, D., Imondi, A.R., Placke, M.E., Vail, D.M., and Kris, M.G. (2001) Development of inhalational agents for oncologic use. *J Clin Oncol*, **19**, 1839–1847.
4. Mendelsohn, J., and Baselga, J. (2000) The EGF receptor family as targets for cancer therapy. *Oncogene*, **19**, 6550–6565.
5. Mukohara, T., Engelman, J.A., Hanna, N.H., Yeap, B.Y., Kobayashi, S., Lindeman, N., Halmos, B., Pearlberg, J., Tsuchihashi, Z., Cantley, L.C., Tenen, D.G., Johnson, B.E., and Janne, P.A. (2005) Differential effects of gefitinib and cetuximab on non-small-cell lung cancers bearing epidermal growth factor receptor mutations. *J Natl Cancer Inst*, **97**, 1185–1194.
6. Reade, C.A., and Ganti, A.K. (2009) EGFR targeted therapy in non-small cell lung cancer: potential role of cetuximab. *Biologics*, **3**, 215–224.
7. Takeuchi, K., and Ito, F. (2010) EGF receptor in relation to tumor development: molecular basis of responsiveness of cancer cells to EGFR-targeting tyrosine kinase inhibitors. *FEBS J*, **277**, 316–326.
8. Campbell, L., Blackhall, F., and Thatcher, N. (2010) Gefitinib for the treatment of non-small-cell lung cancer. *Expert Opin Pharmacother*, **11**, 1343–1357.
9. Iyer, R., and Bharthuar, A. (2010) A review of erlotinib--an oral, selective epidermal growth factor receptor tyrosine kinase inhibitor. *Expert Opin Pharmacother*, **11**, 311–320.
10. Okamoto, I. (2010) Epidermal growth factor receptor in relation to tumor development: EGFR-targeted anticancer therapy. *FEBS J*, **277**, 309–315.
11. Gerber, D.E., and Choy, H. (2010) Cetuximab in combination therapy: from bench to clinic. *Cancer Metastasis Rev*, **29**, 171–180.
12. Jain, K.K. (2008) Recent advances in nanooncology. *Technol Cancer Res Treat*, **7**, 1–13.
13. Alexis, F., Rhee, J.W., Richie, J.P., Radovic-Moreno, A.F., Langer, R., and Farokhzad, O.C. (2008) New frontiers in nanotechnology for cancer treatment. *Urol Oncol*, **26**, 74–85.

14. Haley, B., and Frenkel, E. (2008) Nanoparticles for drug delivery in cancer treatment. *Urol Oncol*, **26**, 57–64.
15. Jiang, W., Kim, B.Y., Rutka, J.T., and Chan, W.C. (2008) Nanoparticle-mediated cellular response is size-dependent. *Nat Nanotechnol*, **3**, 145–150.
16. Peer, D., Karp, J.M., Hong, S., Farokhzad, O.C., Margalit, R., and Langer, R. (2007) Nanocarriers as an emerging platform for cancer therapy. *Nat Nanotechnol*, **2**, 751–760.
17. Ruoslahti, E., Bhatia, S.N., and Sailor, M.J. (2010) Targeting of drugs and nanoparticles to tumors. *J Cell Biol*, **188**, 759–768.
18. Matsumura, Y., and Maeda, H. (1986) A new concept for macromolecular therapeutics in cancer chemotherapy: mechanism of tumoritropic accumulation of proteins and the antitumor agent smancs. *Cancer Res*, **46**, 6387–6392.
19. Tseng, C.L., Wu, S.Y., Wang, W.H., Peng, C.L., Lin, F.H., Lin, C.C., Young, T.H., and Shieh, M.J. (2008) Targeting efficiency and biodistribution of biotinylated-EGF-conjugated gelatin nanoparticles administered via aerosol delivery in nude mice with lung cancer. *Biomaterials*, **29**, 3014–3022.
20. Colombo, M., Corsi, F., Foschi, D., Mazzantini, E., Mazzucchelli, S., Morasso, C., Occhipinti, E., Polito, L., Prosperi, D., Ronchi, S., and Verderio, P. (2010) HER2 targeting as a two-sided strategy for breast cancer diagnosis and treatment: outlook and recent implications in nanomedical approaches. *Pharmacol Res*, **62**, 150–165.
21. Zhao, X., Li, H., and Lee, R.J. (2008) Targeted drug delivery via folate receptors. *Expert Opin Drug Deliv*, **5**, 309–319.
22. Choi, C.H., Alabi, C.A., Webster, P., and Davis, M.E. (2010) Mechanism of active targeting in solid tumors with transferrin-containing gold nanoparticles. *Proc Natl Acad Sci USA*, **107**, 1235–1240.
23. Arap, W., Pasqualini, R., and Ruoslahti, E. (1998) Cancer treatment by targeted drug delivery to tumor vasculature in a mouse model. *Science*, **279**, 377–380.
24. Porkka, K., Laakkonen, P., Hoffman, J.A., Bernasconi, M., and Ruoslahti, E. (2002) A fragment of the HMGN2 protein homes to the nuclei of tumor cells and tumor endothelial cells in vivo. *Proc Natl Acad Sci USA*, **99**, 7444–7449.
25. Curnis, F., Gasparri, A., Sacchi, A., Longhi, R., and Corti, A. (2004) Coupling tumor necrosis factor-alpha with alphaV integrin ligands improves its antineoplastic activity. *Cancer Res*, **64**, 565–571.

26. Paoloni, M.C., Tandle, A., Mazcko, C., Hanna, E., Kachala, S., Leblanc, A., Newman, S., Vail, D., Henry, C., Thamm, D., Sorenmo, K., Hajitou, A., Pasqualini, R., Arap, W., Khanna, C., and Libutti, S.K. (2009) Launching a novel preclinical infrastructure: comparative oncology trials consortium directed therapeutic targeting of TNFalpha to cancer vasculature. *PLoS One*, **4**, e4972.

27. Peer, D., and Margalit, R. (2004) Tumor-targeted hyaluronan nanoliposomes increase the antitumor activity of liposomal Doxorubicin in syngeneic and human xenograft mouse tumor models. *Neoplasia*, **6**, 343–353.

28. Peer, D., and Margalit, R. (2004) Loading mitomycin C inside long circulating hyaluronan targeted nano-liposomes increases its antitumor activity in three mice tumor models. *Int J Cancer*, **108**, 780–789.

29. Loebinger, M.R., Kyrtatos, P.G., Turmaine, M., Price, A.N., Pankhurst, Q., Lythgoe, M.F., and Janes, S.M. (2009) Magnetic resonance imaging of mesenchymal stem cells homing to pulmonary metastases using biocompatible magnetic nanoparticles. *Cancer Res*, **69**, 8862–8867.

30. Tseng, C.L., Su, W.Y., Yen, K.C., Yang, K.C., and Lin, F.H. (2009) The use of biotinylated-EGF-modified gelatin nanoparticle carrier to enhance cisplatin accumulation in cancerous lungs via inhalation. *Biomaterials*, **30**, 3476–3485.

31. Surendiran, A., Sandhiya, S., Pradhan, S.C., and Adithan, C. (2009) Novel applications of nanotechnology in medicine. *Indian J Med Res*, **130**, 689–701.

32. Gradishar, W.J., Tjulandin, S., Davidson, N., Shaw, H., Desai, N., Bhar, P., Hawkins, M., and O'Shaughnessy, J. (2005) Phase III trial of nanoparticle albumin-bound paclitaxel compared with polyethylated castor oil-based paclitaxel in women with breast cancer. *J Clin Oncol*, **23**, 7794–7803.

33. Reynolds, C., Barrera, D., Jotte, R., Spira, A.I., Weissman, C., Boehm, K.A., Pritchard, S., and Asmar, L. (2009) Phase II trial of nanoparticle albumin-bound paclitaxel, carboplatin, and bevacizumab in first-line patients with advanced nonsquamous non-small cell lung cancer. *J Thorac Oncol*, **4**, 1537–1543.

34. Ramesh, R., Saeki, T., Templeton, N.S., Ji, L., Stephens, L.C., Ito, I., Wilson, D.R., Wu, Z., Branch, C.D., Minna, J.D., and Roth, J.A. (2001) Successful treatment of primary and disseminated human lung cancers by systemic delivery of tumor suppressor genes using an improved liposome vector. *Mol Ther*, **3**, 337–350.

35. Ito, I., Ji, L., Tanaka, F., Saito, Y., Gopalan, B., Branch, C.D., Xu, K., Atkinson, E.N., Bekele, B.N., Stephens, L.C., Minna, J.D., Roth, J.A., and Ramesh, R. (2004) Liposomal vector mediated delivery of the 3p FUS1 gene demonstrates potent antitumor activity against human lung cancer in vivo. *Cancer Gene Ther*, **11**, 733–739.

36. Gopalan, B., Ito, I., Branch, C.D., Stephens, C., Roth, J.A., and Ramesh, R. (2004) Nanoparticle based systemic gene therapy for lung cancer: molecular mechanisms and strategies to suppress nanoparticle-mediated inflammatory response. *Technol Cancer Res Treat*, **3**, 647–657.

37. Ramesh, R., Ito, I., Saito, Y., Wu, Z., Mhashikar, A.M., Wilson, D.R., Branch, C.D., Roth, J.A., and Chada, S. (2004) Local and systemic inhibition of lung tumor growth after nanoparticle-mediated mda-7/IL-24 gene delivery. *DNA Cell Biol*, **23**, 850–857.

38. Templeton, N.S., Lasic, D.D., Frederik, P.M., Strey, H.H., Roberts, D.D., and Pavlakis, G.N. (1997) Improved DNA: liposome complexes for increased systemic delivery and gene expression. *Nat Biotechnol*, **15**, 647–652.

39. Ramesh, R. (2008) Nanoparticle-mediated gene delivery to the lung. *Methods Mol Biol*, **433**, 301–331.

40. Mccormack, B., and Gregoriadis, G. (1994) Entrapment of cyclodextrin drug complexes into liposomes–potential advantages in drug-delivery. *J Drug Targeting*, **2**, 449–454.

41. Desai, T.A., Chu, W.H., Tu, J.K., Beattie, G.M., Hayek, A., and Ferrari, M. (1998) Microfabricated immunoisolating biocapsules. *Biotechnol Bioeng*, **57**, 118–120.

42. Freitas, R.A. (2005) Current status of nanomedicine and medical nanorobotics. *J Comput Theor Nanosci*, **2**, 1–25.

43. Kroto, H.W., Heath, J.R., Obrien, S.C., Curl, R.F., and Smalley, R.E. (1985) C-60–Buckminsterfullerene. *Nature*, **318**, 162–163.

44. Mroz, P., Pawlak, A., Satti, M., Lee, H., Wharton, T., Gali, H., Sarna, T., and Hamblin, M.R. (2007) Functionalized fullerenes mediate photodynamic killing of cancer cells: Type I versus Type II photochemical mechanism. *Free Radic Biol Med*, **43**, 711–719.

45. Ji, H., Yang, Z., Jiang, W., Geng, C., Gong, M., Xiao, H., Wang, Z., and Cheng, L. (2008) Antiviral activity of nano carbon fullerene lipidosome against influenza virus in vitro. *J Huazhong Univ Sci Technolog Med Sci*, **28**, 243–246.

46. Markovic, Z., and Trajkovic, V. (2008) Biomedical potential of the reactive oxygen species generation and quenching by fullerenes (C60). *Biomaterials*, **29**, 3561–3573.

47. Iijima, S. (1991) Helical microtubules of graphitic carbon. *Nature*, **354**, 56–58.

48. Prato, M., Kostarelos, K., and Bianco, A. (2008) Functionalized carbon nanotubes in drug design and discovery. *Acc Chem Res*, **41**, 60–68.

49. Zhang, Z., Yang, X., Zhang, Y., Zeng, B., Wang, S., Zhu, T., Roden, R.B., Chen, Y., and Yang, R. (2006) Delivery of telomerase reverse transcriptase small interfering RNA in complex with positively charged single-walled carbon nanotubes suppresses tumor growth. *Clin Cancer Res*, **12**, 4933–4939.

50. Walling, M.A., Novak, J.A., and Shepard, J.R. (2009) Quantum dots for live cell and in vivo imaging. *Int J Mol Sci*, **10**, 441–491.

51. Gao, X., Cui, Y., Levenson, R.M., Chung, L.W., and Nie, S. (2004) In vivo cancer targeting and imaging with semiconductor quantum dots. *Nat Biotechnol*, **22**, 969–976.

52. Iga, A.M., Robertson, J.H., Winslet, M.C., and Seifalian, A.M. (2007) Clinical potential of quantum dots. *J Biomed Biotechnol*, **2007**, 76087.

53. West, J.L., and Halas, N.J. (2000) Applications of nanotechnology to biotechnology commentary. *Curr Opin Biotechnol*, **11**, 215–217.

54. Hirsch, L.R., Stafford, R.J., Bankson, J.A., Sershen, S.R., Rivera, B., Price, R.E., Hazle, J.D., Halas, N.J., and West, J.L. (2003) Nanoshell-mediated near-infrared thermal therapy of tumors under magnetic resonance guidance. *Proc Natl Acad Sci USA*, **100**, 13549–13554.

55. Choi, M.R., Stanton-Maxey, K.J., Stanley, J.K., Levin, C.S., Bardhan, R., Akin, D., Badve, S., Sturgis, J., Robinson, J.P., Bashir, R., Halas, N.J., and Clare, S.E. (2007) A cellular Trojan Horse for delivery of therapeutic nanoparticles into tumors. *Nano Lett*, **7**, 3759–3765.

56. Gao, Z., Kennedy, A.M., Christensen, D.A., and Rapoport, N.Y. (2008) Drug-loaded nano/microbubbles for combining ultrasonography and targeted chemotherapy. *Ultrasonics*, **48**, 260–270.

57. Rapoport, N., Gao, Z., and Kennedy, A. (2007) Multifunctional nanoparticles for combining ultrasonic tumor imaging and targeted chemotherapy. *J Natl Cancer Inst*, **99**, 1095–1106.

58. Artemov, D., Mori, N., Okollie, B., and Bhujwalla, Z.M. (2003) MR molecular imaging of the Her-2/neu receptor in breast cancer cells using targeted iron oxide nanoparticles. *Magn Reson Med*, **49**, 403–408.

59. Leuschner, C., Kumar, C.S., Hansel, W., Soboyejo, W., Zhou, J., and Hormes, J. (2006) LHRH-conjugated magnetic iron oxide nanoparticles for detection of breast cancer metastases. *Breast Cancer Res Treat*, **99**, 163–176.

60. Cuenca, A.G., Jiang, H., Hochwald, S.N., Delano, M., Cance, W.G., and Grobmyer, S.R. (2006) Emerging implications of nanotechnology on cancer diagnostics and therapeutics. *Cancer*, **107**, 459–466.

61. Weissleder, R., Elizondo, G., Wittenberg, J., Lee, A.S., Josephson, L., and Brady, T.J. (1990) Ultrasmall superparamagnetic iron oxide: an intravenous contrast agent for assessing lymph nodes with MR imaging. *Radiology*, **175**, 494–498.

62. Li, Z., Kawashita, M., Araki, N., Mitsumori, M., Hiraoka, M., and Doi, M. (2010) Preparation of magnetic iron oxide nanoparticles for hyperthermia of cancer in a $FeCl_2$-$NaNO_3$-NaOH aqueous system. *J Biomater Appl*, **25**, 643–641.

63. Mottram, P.L. (2003) Past, present and future drug treatment for rheumatoid arthritis and systemic lupus erythematosus. *Immunol Cell Biol*, **81**, 350–353.

64. Sokolov, K., Follen, M., Aaron, J., Pavlova, I., Malpica, A., Lotan, R., and Richards-Kortum, R. (2003) Real-time vital optical imaging of precancer using anti-epidermal growth factor receptor antibodies conjugated to gold nanoparticles. *Cancer Res*, **63**, 1999–2004.

65. Paciotti, G.F., Myer, L., Weinreich, D., Goia, D., Pavel, N., McLaughlin, R.E., and Tamarkin, L. (2004) Colloidal gold: a novel nanoparticle vector for tumor directed drug delivery. *Drug Deliv*, **11**, 169–183.

66. Powell, A.C., Paciotti, G.F., and Libutti, S.K. (2010) Colloidal gold: a novel nanoparticle for targeted cancer therapeutics. *Methods Mol Biol*, **624**, 375–384.

67. Hainfeld, J.F., Slatkin, D.N., and Smilowitz, H.M. (2004) The use of gold nanoparticles to enhance radiotherapy in mice. *Phys Med Biol*, **49**, N309–N315.

68. Ferrari, M. (2008) Nanogeometry: beyond drug delivery. *Nat Nanotechnol*, **3**, 131–132.

69. Duguet, E., Vasseur, S., Mornet, S., and Devoisselle, J.M. (2006) Magnetic nanoparticles and their applications in medicine. *Nanomedicine (London)*, **1**, 157–168.

70. Shubayev, V.I., Pisanic, T.R., 2nd, and Jin, S. (2009) Magnetic nanoparticles for theragnostics. *Adv Drug Deliv Rev*, **61**, 467–477.

71. Perrault, S.D., Walkey, C., Jennings, T., Fischer, H.C., and Chan, W.C. (2009) Mediating tumor targeting efficiency of nanoparticles through design. *Nano Lett*, **9**, 1909–1915.

72. Gupta, A.K., and Gupta, M. (2005) Synthesis and surface engineering of iron oxide nanoparticles for biomedical applications. *Biomaterials*, **26**, 3995–4021.

73. Dobrovolskaia, M.A., and McNeil, S.E. (2007) Immunological properties of engineered nanomaterials. *Nat Nanotechnol*, **2**, 469–478.

74. Yamazaki, M., and Ito, T. (1990) Deformation and instability in membrane structure of phospholipid vesicles caused by osmophobic association: mechanical stress model for the mechanism of poly(ethylene glycol)-induced membrane fusion. *Biochemistry*, **29**, 1309–1314.

75. Renwick, L.C., Brown, D., Clouter, A., and Donaldson, K. (2004) Increased inflammation and altered macrophage chemotactic responses caused by two ultrafine particle types. *Occup Environ Med*, **61**, 442–447.

76. Labiris, N.R., and Dolovich, M.B. (2003) Pulmonary drug delivery. Part I: physiological factors affecting therapeutic effectiveness of aerosolized medications. *Br J Clin Pharmacol*, **56**, 588–599.

77. Tatsumura, T., Koyama, S., Tsujimoto, M., Kitagawa, M., and Kagamimori, S. (1993) Further study of nebulization chemotherapy, a new chemotherapeutic method in the treatment of lung carcinomas–fundamental and clinical. *Br J Cancer*, **68**, 1146–1149.

78. Verschraegen, C.F., Gilbert, B.E., Loyer, E., Huaringa, A., Walsh, G., Newman, R.A., and Knight, V. (2004) Clinical evaluation of the delivery and safety of aerosolized liposomal 9-nitro-20(s)-camptothecin in patients with advanced pulmonary malignancies. *Clin Cancer Res*, **10**, 2319–2326.

79. Otterson, G.A., Villalona-Calero, M.A., Sharma, S., Kris, M.G., Imondi, A., Gerber, M., White, D.A., Ratain, M.J., Schiller, J.H., Sandler, A., Kraut, M., Mani, S., and Murren, J.R. (2007) Phase I study of inhaled Doxorubicin for patients with metastatic tumors to the lungs. *Clin Cancer Res*, **13**, 1246–1252.

80. Wittgen, B.P., Kunst, P.W., van der Born, K., van Wijk, A.W., Perkins, W., Pilkiewicz, F.G., Perez-Soler, R., Nicholson, S., Peters, G.J., and Postmus, P.E. (2007) Phase I study of aerosolized SLIT cisplatin in the treatment of patients with carcinoma of the lung. *Clin Cancer Res*, **13**, 2414–2421.

81. Gagnadoux, F., Hureaux, J., Vecellio, L., Urban, T., Le Pape, A., Valo, I., Montharu, J., Leblond, V., Boisdron-Celle, M., Lerondel, S., Majoral, C., Diot, P., Racineux, J.L., and Lemarie, E. (2008) Aerosolized chemotherapy. *J Aerosol Med*, **21**, 61–70.

82. Karlsson, H.L., Cronholm, P., Gustafsson, J., and Moller, L. (2008) Copper oxide nanoparticles are highly toxic: a comparison between metal oxide nanoparticles and carbon nanotubes. *Chem Res Toxicol*, **21**, 1726–1732.

83. Abe, S., Takizawa, H., Sugawara, I., and Kudoh, S. (2000) Diesel exhaust (DE)-induced cytokine expression in human bronchial epithelial cells: a study with a new cell exposure system to freshly generated DE in vitro. *Am J Respir Cell Mol Biol*, **22**, 296–303.

84. Valinluck, V., and Sowers, L.C. (2007) Inflammation-mediated cytosine damage: a mechanistic link between inflammation and the epigenetic alterations in human cancers. *Cancer Res*, **67**, 5583–5586.

85. Federico, A., Morgillo, F., Tuccillo, C., Ciardiello, F., and Loguercio, C. (2007) Chronic inflammation and oxidative stress in human carcinogenesis. *Int J Cancer*, **121**, 2381–2386.

86. Shukla, R., Bansal, V., Chaudhary, M., Basu, A., Bhonde, R.R., and Sastry, M. (2005) Biocompatibility of gold nanoparticles and their endocytotic fate inside the cellular compartment: a microscopic overview. *Langmuir*, **21**, 10644–10654.

87. Li, J.J., Zou, L., Hartono, D., Ong, C.N., Bay, B.H., and Yung, L.Y.L. (2008) Gold nanoparticles induce oxidative damage in lung fibroblasts in vitro. *Adv Mater*, **20**, 138–+.

88. Stevens, R.G., Jones, D.Y., Micozzi, M.S., and Taylor, P.R. (1988) Body iron stores and the risk of cancer. *N Engl J Med*, **319**, 1047–1052.

89. Toyokuni, S. (2002) Iron and carcinogenesis: from Fenton reaction to target genes. *Redox Rep*, **7**, 189–197.

90. Bhasin, G., Kauser, H., and Athar, M. (2002) Iron augments stage-I and stage-II tumor promotion in murine skin. *Cancer Lett*, **183**, 113–122.

91. Sayes, C.M., Reed, K.L., and Warheit, D.B. (2007) Assessing toxicity of fine and nanoparticles: comparing in vitro measurements to in vivo pulmonary toxicity profiles. *Toxicol Sci*, **97**, 163–180.

92. Fischer, H.C., and Chan, W.C. (2007) Nanotoxicity: the growing need for in vivo study. *Curr Opin Biotechnol*, **18**, 565–71.

93. Sokolov, K., Tam, J., Travis, K., Larson, T., Aaron, J., Harrison, N., Emelianov, S., and Johnston, K. (2009) Cancer imaging and therapy with metal nanoparticles. *Conf Proc IEEE Eng Med Biol Soc*, **2009**, 2005–2007.

94. Aaron, J.S., Oh, J., Larson, T.A., Kumar, S., Milner, T.E., and Sokolov, K.V. (2006) Increased optical contrast in imaging of epidermal growth

factor receptor using magnetically actuated hybrid gold/iron oxide nanoparticles. *Opt Express*, **14**, 12930–12943.

95. Goldstein, N.I., Prewett, M., Zuklys, K., Rockwell, P., and Mendelsohn, J. (1995) Biological efficacy of a chimeric antibody to the epidermal growth factor receptor in a human tumor xenograft model. *Clin Cancer Res*, **1**, 1311–1318.

96. Doody, J.F., Wang, Y., Patel, S.N., Joynes, C., Lee, S.P., Gerlak, J., Rolser, R.L., Li, Y., Steiner, P., Bassi, R., Hicklin, D.J., and Hadari, Y.R. (2007) Inhibitory activity of cetuximab on epidermal growth factor receptor mutations in non small cell lung cancers. *Mol Cancer Ther*, **6**, 2642–2651.

97. Rosell, R., Robinet, G., Szczesna, A., Ramlau, R., Constenla, M., Mennecier, B.C., Pfeifer, W., O'Byrne, K.J., Welte, T., Kolb, R., Pirker, R., Chemaissani, A., Perol, M., Ranson, M.R., Ellis, P.A., Pilz, K., and Reck, M. (2008) Randomized phase II study of cetuximab plus cisplatin/vinorelbine compared with cisplatin/vinorelbine alone as first-line therapy in EGFR-expressing advanced non-small-cell lung cancer. *Ann Oncol*, **19**, 362–369.

98. Hanna, N., Lilenbaum, R., Ansari, R., Lynch, T., Govindan, R., Janne, P.A., and Bonomi, P. (2006) Phase II trial of cetuximab in patients with previously treated non-small-cell lung cancer. *J Clin Oncol*, **24**, 5253–5258.

99. Hughes, S., Liong, J., Miah, A., Ahmad, S., Leslie, M., Harper, P., Prendiville, J., Shamash, J., Subramaniam, R., Gaya, A., Spicer, J., and Landau, D. (2008) A brief report on the safety study of induction chemotherapy followed by synchronous radiotherapy and cetuximab in stage III non-small cell lung cancer (NSCLC): SCRATCH study. *J Thorac Oncol*, **3**, 648–651.

100. Ma, L.L., Tam, J.O., Willsey, B.W., Rigdon, D., Ramesh, R., Sokolov, K., and Johnston, K.P. (2011) Selective targeting of antibody conjugated multifunctional nanoclusters (nanoroses) to epidermal growth factor receptors in cancer cells. *Langmuir*, **27**, 7681–7690.

101. Yokoyama, T., Tam, J., Kuroda, S., Scott, A.W., Aaron, J., Larson, T., Shanker, M., Correa, A.M., Kondo, S., Roth, J.A., Sokolov, K., and Ramesh, R. (2011) EGFR-targeted hybrid plasmonic magnetic nanoparticles synergistically induce autophagy and apoptosis in non-small cell lung cancer cells. *PLoS One*, **6**, e25507.

102. Waksal, H.W. (1999) Role of an anti-epidermal growth factor receptor in treating cancer. *Cancer Metastasis Rev*, **18**, 427–436.

103. Singh, N., Manshian, B., Jenkins, G.J., Griffiths, S.M., Williams, P.M., Maffeis, T.G., Wright, C.J., and Doak, S.H. (2009) NanoGenotoxicology:

the DNA damaging potential of engineered nanomaterials. *Biomaterials*, **30**, 3891–3914.

104. Ahamed, M., Karns, M., Goodson, M., Rowe, J., Hussain, S.M., Schlager, J.J., and Hong, Y. (2008) DNA damage response to different surface chemistry of silver nanoparticles in mammalian cells. *Toxicol Appl Pharmacol*, **233**, 404–410.

105. Mroz, R.M., Schins, R.P., Li, H., Jimenez, L.A., Drost, E.M., Holownia, A., MacNee, W., and Donaldson, K. (2008) Nanoparticle-driven DNA damage mimics irradiation-related carcinogenesis pathways. *Eur Respir J*, **31**, 241–251.

106. Kastan, M.B., and Bartek, J. (2004) Cell-cycle checkpoints and cancer. *Nature*, **432**, 316–323.

107. Ting, N.S., and Lee, W.H. (2004) The DNA double-strand break response pathway: becoming more BRCAish than ever. *DNA Repair (Amsterdam)*, **3**, 935–944.

108. Sokolov, K., Nida, D., Descour, M., Lacy, A., Levy, M., Hall, B., Dharmawardhane, S., Ellington, A., Korgel, B., and Richards-Kortum, R. (2007) Molecular optical imaging of therapeutic targets of cancer. *Adv Cancer Res*, **96**, 299–344.

109. Lu, M., Cohen, M.H., Rieves, D., and Pazdur, R. (2010) FDA report: Ferumoxytol for intravenous iron therapy in adult patients with chronic kidney disease. *Am J Hematol*, **85**, 315–319.

Chapter 3

Nasal and Pulmonary Delivery of Macromolecules to Treat Respiratory and Nonrespiratory Diseases

Durga Paturi, Mitesh Patel, Ranjana Mitra, and Ashim K. Mitra*

Division of Pharmaceutical Sciences, School of Pharmacy, University of Missouri-Kansas City, 2464 Charlotte Street, Kansas City, MO 64108-2718, US

*mitraa@umkc.edu

3.1 Introduction

The delivery of macromolecules such as peptides, proteins, vaccines, and genes has gained considerable attention in recent years. Macromolecules have been the drugs of choice due to their high specificity and selectivity. However, the delivery of such compounds suffers from several limitations such as a larger size, poor solubility, rapid degradation in the biological environment, poor bioavailability, and a higher cost of manufacture. Furthermore, one of the most difficult tasks for scientists today has been the delivery of these macromolecules to their site of action [1, 2]. Pain due to injection and patient compliance led to the exploration of alternative routes such as nasal, transdermal, and buccal. The respiratory route is one such potential route for alternative, noninvasive delivery. Macromolecules can be administered either to the upper respiratory route via the nasal cavity or the lower respiratory route via the lung. Recent advances in biotechnology, inhaler devices, absorption enhancers, and targeting motifs have led to a substantial increase in

Pulmonary Nanomedicine: Diagnostics, Imaging, and Therapeutics
Edited by Neeraj Vij

ISBN 978-981-4316-48-4 (Hardback), 978-981-4364-14-0 (eBook)
www.panstanford.com

research in these areas. The development of protein and nucleic acid combinatorial libraries allowed the generation of lead molecules that facilitated the targeted delivery of macromolecules. In this book chapter, the nasal and pulmonary route and the absorption mechanisms across the mucosal membranes will be discussed. This chapter will also focus on the various particulate delivery systems utilized for the formulation of macromolecules.

3.2 Nasal Drug Delivery

Nasal delivery offers significant advantages over other noninvasive routes. These include ease of self-administration, accurate dosing, a highly vascularized mucosa leading to rapid absorption, and high systemic bioavailability. This delivery route offers a much lower enzyme activity relative to the oral mucosa and also avoids the first-pass metabolism associated with the oral route. The extent of macromolecular drug delivery across the nasal mucosa is inversely proportional to the molecular weight. Therefore, in order to increase nasal bioavailability, new approaches have been developed, which include penetration enhancers and the development of prodrugs, analogues, and particulate systems composed of biodegradable polymers. Absorption through the nasal mucosa is primarily dependent on the molecular size and hydrophobicity of the molecules [3, 4]. Low molecular weight hydrophilic molecules permeate readily through the porous aqueous channels of the endothelial basement membrane of the nasal epithelium. However, highly polar macromolecules encounter high resistance in passing through these channels. Therefore, new approaches should be investigated to enhance nasal permeability of polar, high molecular weight therapeutic agents, particularly for proteins and vaccines intended to elicit a systemic response.

Nasal epithelial cells also express various transporters and receptors for macromolecular entry. Moreover, macromolecules as well as particulate polymeric nano- and microparticles can penetrate the epithelial cell layer by phagocytosis, micropinocytosis, and endocytosis by receptor-mediated or nonreceptor-mediated vesicular pathways. Viral vectors such as adeno-associated virus (AAV), cytomegalovirus (CMV), herpes simplex virus (HSV), and retrovirus (RV) could be engineered to deliver genes and other macromolecules across the nasal epithelial cells. Liposomes and nanoparticles can be surface decorated with the outer protein coat of the virus to enhance cellular entry.

3.2.1 Nasal Anatomy

The nose provides the sense of smell and also functions to filter, humidify, entrap, and clear particles larger than 10 μm in size. The outer part of the nose is composed of bones and cartilage (Fig. 3.1A). The nasal cavity is separated into left and right passages by a nasal septum and is formed by cranial bones, the nasal bone, the ethmoid bone, and the vomer bone. These bones branch into highly folded segments called turbinates, which project along the lateral walls of the nasal passages. These interwined structures enhance the contact surface area between the inhaled air and the nasal epithelium and

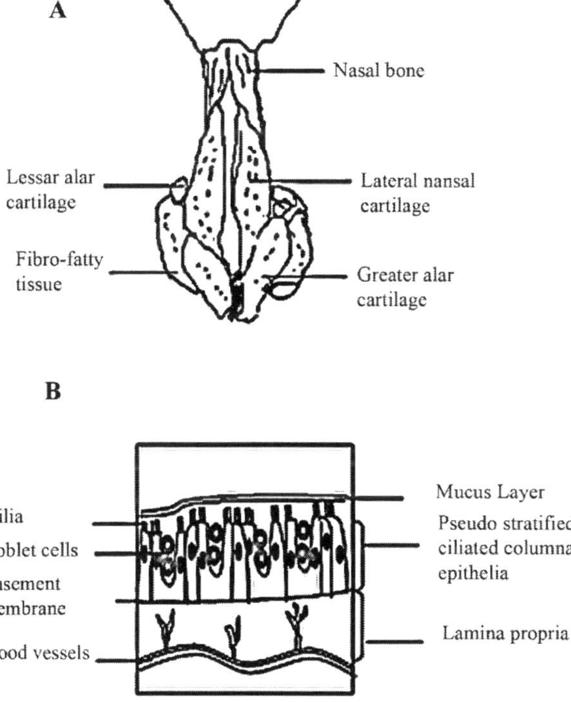

Figure 3.1 (A) The anatomy of the nose. (B) The nasal mucosa.

act as a barrier for the foreign particles entering from the lower respiratory track. Air enters the nasal cavity through the two openings formed by the nasal cartilage and nostrils. Each of the nasal cavities is 2–4 mm wide and is lined with three different functional regions: vestibular, respiratory, and olfactory.

The majority of the nasal cavity is lined by the respiratory epithelium. The stratified squamous epithelium lines the anterior one-third of the nasal cavity. The posterior of the nasal cavity is lined with the pseudostratified ciliated columnar epithelium interspersed with goblet cells (Fig. 3.1B). Both, the nostrils and the nasal cavity have a lining of mucous membrane and cilia. The mucous membrane lining the nasal airways has two layers — the luminal surface epithelium and underlying connective tissue lamina propria. This connective tissue contains various types of blood vessels, lymphatic vessels, nerves, glands, and mesenchymal cells that are embedded in the matrix. The luminal surface of the epithelium is covered by mucus secreted by goblet cells. Mucus plays a role in the upper airway defense mechanism filtering inhaled air and entrapping the foreign particles. The mucus, with the entrapped particles, is propelled by the cilia present on the surface of the epithelium. The nasal mucociliary apparatus provides the first line defense against inhaled pathogens [5, 6].

3.2.2 Mechanisms of Nasal Absorption

Macromolecules have a short half-life due to which they require repeated administration. Molecules that are eliminated by first pass metabolism can directly enter the systemic circulation by intranasal route. Ciliated columnar cells comprise most of the respiratory region of the nose. Columnar cells in the respiratory region are covered by approximately 300 microvilli, which greatly enhance the surface area of the nasal cavity. Nasal mucosa is highly vascularized, and the capillary permeability has been reported to be much greater than that of intestine. The mucosa is covered with layers of mucus, and, therefore, it is essential that the macromolecular drugs first permeate through these layers. Once macromolecular drugs pass through the mucus layers, the molecules can traverse the nasal mucosa in three different ways. These three modes of transport are paracellular transport, transcellular transport, and transcytosis

(Fig. 3.2). Uncharged molecules with low molecular weights can easily pass through the mucus layer whereas large and/or charged particles cannot readily permeate this layer [7]. Depending on hydrophilicity or lipophilicity of the macromolecular drugs, these compounds can be either passively transported through the paracellular route or can be actively or passively transported through the transcellular pathway. The paracellular route includes passage of the macromolecular drugs through the intercellular spaces or through the epithelial tight junctions. The opening of the tight junctions improves absorption of hydrophilic macromolecular drugs. It has been reported that chitosan improves drug transport by loosening the epithelial tight junctions [8]. Lipophilic macromolecular drugs undergoing permeation via the transcellular pathway can be either passively diffused or actively transported by carrier systems present on the cell membrane. In transcytosis, the macromolecular drugs are internalized in vesicles that eventually accumulate in the cytoplasm [9].

Lang *et al.* mathematically derived the effective permeability coefficient under steady-state conditions using thymocartin across excised bovine nasal mucosa:

$$P_{\text{eff}} = (dc/dt)\text{ss}\, V/\,(A\ CD) \tag{3.1}$$

where, (dc/dt)ss represents the change in concentration with time at steady state, A is the permeation area, V is the volume of the receiver compartment, and CD is the initial concentration in the donor compartment [10].

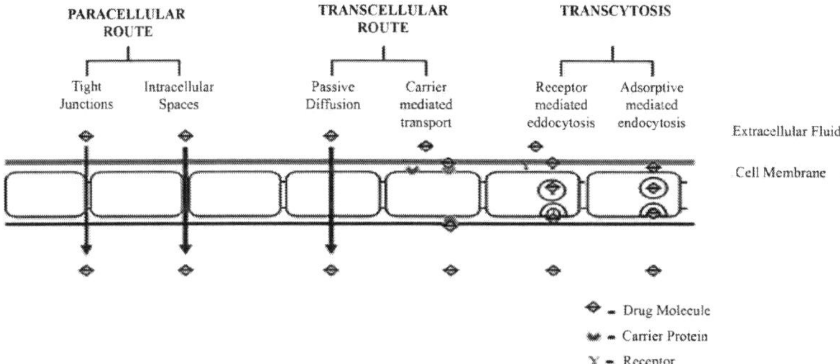

Figure 3.2 Various mechanisms for the transport of proteins and peptide drugs across the nasal mucosa.

3.2.3 Factors Affecting Nasal Absorption

Several factors can affect the absorption of macromolecular drugs across the nasal mucosa. These include physiological factors, pathological conditions, biochemical changes, physicochemical properties of the drug, physicochemical properties of the formulation, drug distribution, dosage forms, and factors related to delivery devices (Fig. 3.3 on next page) [11].

3.2.3.1 Physiological factors

3.2.3.1.1 *Blood supply*

The nasal mucosa is highly vascularized (venous sinusoids and arteriovenous anastomosis) and has an enormous surface area (of 150 cm^2) that enhances nasal absorption.

3.2.3.1.2 *Nasal secretion*

Serous and seromucous glands produce nasal secretions. A total of 1.5 to 21 mL of mucus is produced every day, which forms a double layer of 5 μm thick [12].

3.2.3.1.3 *Mucociliary clearance and cilia beating*

A decrease in mucociliary clearance enhances the contact time of the drug to the mucosa, resulting in improved drug absorption, while an increase in the mucociliary clearance might reduce drug absorption. Each epithelial cell on the nasal mucosa has around 300 hair and like structures known as cilia. The mucus layer and cilia together represent a fluctuating movement, which accelerates drug clearance with a half-life of 21 minutes from the nasal cavity [13].

3.2.3.2 Pathological conditions

Pathological conditions such as common cold, rhinitis, atropic rhinitis, and nasal polyposis can alter drug absorption by accelerating their rates of nasal clearance. Severe nasal polyposis can cause nasal obstruction, which can further decrease nasal absorption. Allergy to some irritants can cause extensive nasal secretions, which can, in turn, drain administered drugs prior to absorption [14].

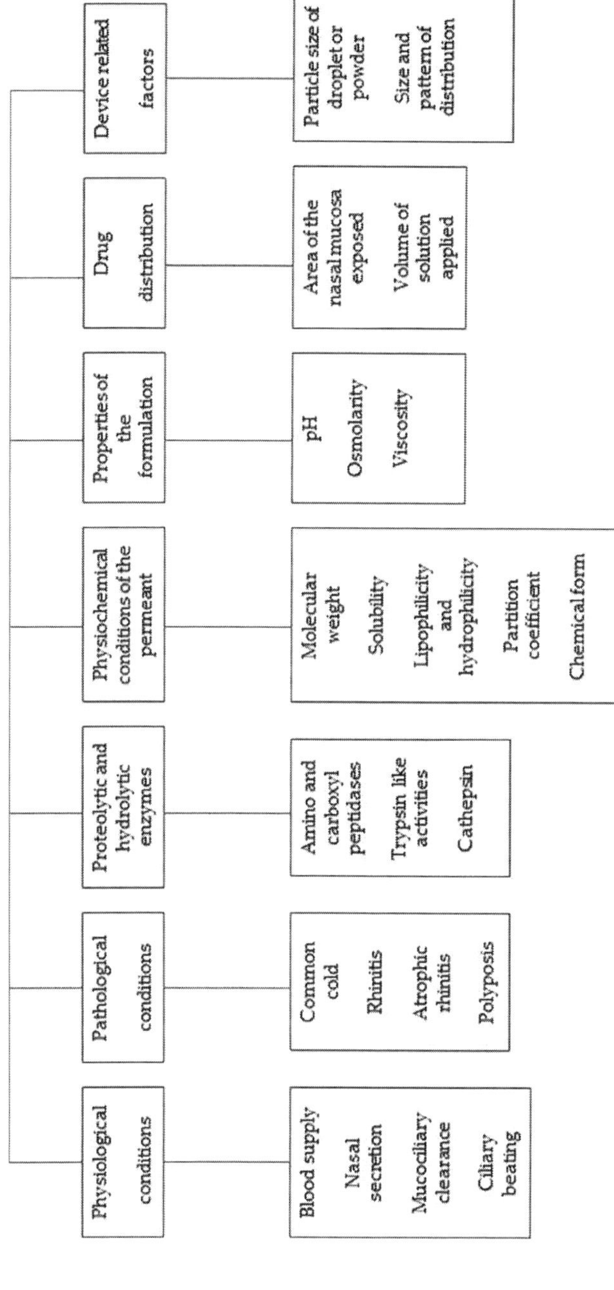

Figure 3.3 Factors affecting nasal absorption.

3.2.3.3 Biochemical changes

The nasal mucosa consists of a number of proteolytic and hydrolytic enzymes such as aminopeptidases, carboxypeptidases, enzymes with trypsin and like activities, and cathepsin that can easily degrade peptide and protein drugs. Out of these enzymes, aminopeptidases are abundantly present in the nasal cavity. Furthermore, the cytochrome P450 activity in the olfactory region is much higher than that in the liver [15].

3.2.3.4 Physicochemical properties of the permeant

Various physicochemical properties of drugs, such as their molecular weights, solubility, lipophilicity and hydrophilicity, and partition coefficients have a significant impact on the rate and extent of absorption across the nasal mucosa.

3.2.3.4.1 *Molecular weight*

It has been reported that the permeability of agents with a molecular weight less than 300 Da is independent of their physicochemical properties and that these agents can easily pass through the aqueous pores or channels of the membranes. On the other hand, the permeability of the drugs with a molecular weight greater than 300 Da is highly dependent on their molecular size. Further, nasal absorption decreases significantly for drugs above a molecular weight of 1,000 Da [4]. The bioavailability of most of the peptide and protein drugs of molecular weights greater than 1 kDa is dependent on their molecular weights. Usually, these macromolecules have an overall bioavailability within 0.5% to 5% [16].

3.2.3.4.2 *Solubility*

A compound has to dissolve in the nasal secretions so as to permeate through the nasal mucosa. Thus, aqueous solubility can be an important factor for determining absorption across the mucosal membrane. Nasal secretions are aqueous in nature and, therefore, it is important that the drug has the required aqueous solubility. The solubility and dissolution rate of a drug is the rate-determining step for nasal absorption from powder and suspension formulations.

3.2.3.4.3 *Lipophilicity and hydrophilicity*

Owing to the lipophilic nature of the nasal mucosa, the permeability of a drug increases as its lipophilicity rises. A lipophilic drug can easily permeate the nasal mucosa via the transcellular route as these molecules can readily partition into or diffuse through the lipid bilayer to reach the cytoplasm [17]. However, an increase in lipophilicity can compromise the drug's solubility in nasal secretions, and therefore a major portion of an administered drug may be drained into the nasopharynx via mucociliary clearance [18]. Similarly, an increase in hydrophilicity can also decrease the systemic absorption of the drug across the nasal mucosa due to poor absorption.

3.2.3.4.4 *Partition coefficient*

It is well known that the unionized form of a drug is more readily absorbed through the cell membrane than the ionized form. However, the ionized form of a drug can also permeate through the nasal mucosa [19]. Thus, the partition coefficient might not play an important role in determining drug absorption across the nasal mucosa.

3.2.3.4.5 *Chemical form of a drug*

The salt or the ester form of a drug can affect its absorption across the nasal mucosa. Nasal absorption of carboxylic acid esters of L-tyrosine was far better than L-tyrosine itself.

3.2.3.5 Properties of the formulation

3.2.3.5.1 *pH and osmolarity*

Formulation pH can alter nasal drug absorption. A study has been conducted in rats to study the influence of pH on the nasal

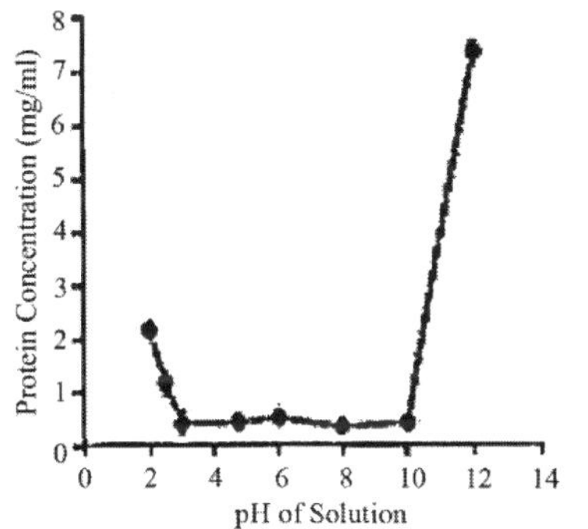

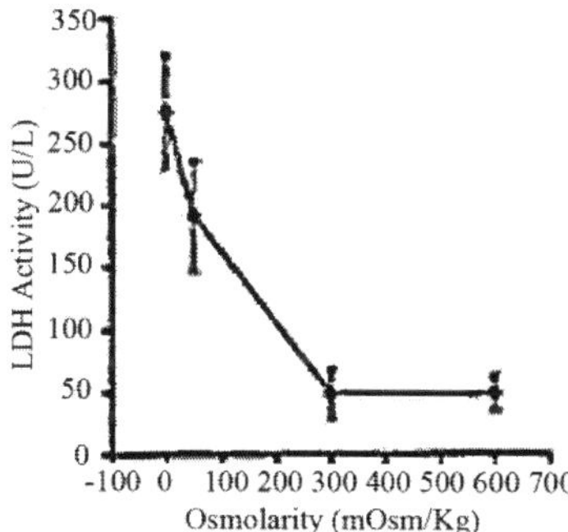

Figure 3.4 Release profiles of protein from the nasal cavity as a function of pH at the end of 120 min perfusion. Phosphate buffers (0.07 M) within pH ranges of 2–12 were used (left). Release profiles of lactate dehydrogenase from the rat nasal cavity as a function of formulation osmolarity (right). Values represent means ≠ SD (n = 3–4).

absorption of secretin. Nasal absorption increased significantly as the pH was reduced from 7.0 to 2.4 [10]. Furthermore, formulations used for intranasal administration should have a pH within 4.5–6.5 so as to prevent mucosal irritation. This article reported that nasal absorption of secretin in rats was maximum with a hyperosmolar saline solution of 0.462 M [20].

Figure 3.4 depicts the effect of solution pH and osmolarity on the nasal epithelial cell integrity using an *in situ* nasal perfusion technique. Solutions within a pH range of 3–10 caused minimal release of total protein and biochemical markers. Also, results from same study indicated that solutions with zero osmolarity caused higher release of lactate dehydrogenase, a cytosolic enzyme [21]

3.2.3.5.2 *Viscosity*

A highly viscous formulation can increase mean residence time at the mucosal surface and prolong mucosal contact time for drug permeation. However, higher viscosity of the formulation may also hinder ciliary beat frequency and mucociliary clearance, thereby altering drug permeation.

3.2.3.6 Drug distribution

The distribution of a drug inside the nasal cavity plays an important role in affecting drug permeation.

3.2.3.6.1 *Area of nasal mucosa exposed*

The influence of nasal mucosal-exposed formulation was investigated by the application of 40 mg progesterone ointment on one nostril as well as on both nostrils of a woman. The results suggested that drug permeation was much higher with application made on both nostrils [4].

3.2.3.6.2 *Volume of the solution applied*

A larger solution volume applied on the nasal cavity can generate a better mucosal distribution profile whereas a small volume might not cover the nasal cavity adequately. The volume that can be applied to the nasal cavity is restricted to 0.05–0.15 mL.

3.2.3.7 Device-related factors

Multiple drug delivery devices are available to administer solution, suspension, and powder formulations into the nasal cavity. The particle size of the droplet or powder and the site and deposition pattern can alter drug absorption profiles.

3.2.3.7.1 *Particle size of droplet and powder*

Particles around 10 μm are deposited in the upper respiratory whereas particles less than 0.5 μm are deposited by the inspired air into the lungs. Particles between 5–7 μm are retained by the nasal cavity.

3.2.3.7.2 *Site and pattern of deposition*

The site and pattern of deposition depend on formulation components, type of the formulation (solution, suspension, and powder), drug delivery devices, design of the actuators or adapters, and the administration technique. The deposition and clearance of Tc-99m-labelled human serum albumin was studied in humans by gamma scintigraphy. The nasal spray is deposited in the anterior part of the nose whereas nasal drops are dispersed in the nasal cavity. Moreover, the nasal drops are more rapidly cleared than the nasal spray [22].

3.2.4 Strategies to Enhance Nasal Absorption

Various strategies may be employed to improve macromolecular drug absorption across the nasal mucosa. A common mechanism includes chelators, which change the physicochemical properties such as drug solubility and partition coefficient, while some enhancers improve absorption by altering nasal mucosal properties. To gain regulatory approval, new penetration enhancers should be nonirritating and nontoxic and should have a transient mild effect on nasal mucosa. The enhancer should produce a transient reversible inhibition of mucosal drug-metabolizing enzymes and prolong contact time of the instilled solutions with the nasal mucosa and reduced mucociliary clearance of the administered particulate dosage forms. Carriers such as liposomes, nanoparticles, microparticles, and gels have been extensively investigated in the past few decades [4]. However, compounds such as cyclodextrins, fusidic acid derivatives, phospholipids, bile salt derivatives, and peptidase or protease inhibitors have been investigated to determine their potential as nasal absorption enhancers.

3.2.4.1 Cyclodextrins

Several cyclodextrins such as α-, β-, and γ-cyclodextrin, methyl cyclodextrin, and hydroxypropyl β-cyclodextrin have been examined for their ability to enhance nasal drug absorption. Among these cyclodextrins, β-cyclodextrin was the only one considered to have a generally recognized as safe (GRAS) status [4]. Merkus *et al.* studied the effects of α-, β-, and γ-cyclodextrins, dimethyl-β-cyclodextrin, and hydroxypropyl-β-cyclodextrin on the intranasal absorption of insulin in rats. This article reported that incorporation of 5% w/v of dimethyl-β-cyclodextrin significantly improved the bioavailability of insulin relative to other cyclodextrins, following intranasal administration of insulin in rats [23].

3.2.4.2 Fusidic acid derivatives

Sodium tauro-24, 25-dihydrofusidate (STDHF) is the most common fusidic acid derivative selected for nasal drug absorption enhancement. It has been employed as a permeation enhancer for the delivery of macromolecules such as insulin, growth hormone, and octreotide [4]. Lee *et al.* studied the bioavailability of intranasally administered salmon calcitonin (sCT) in the presence and absence of 0.5% STDHF in healthy volunteers [24].

3.2.4.3 Phospholipids

Phosphatidylcholines are the primary components of biological membranes. These surface-active amphiphilic phospholipids have been extensively studied as membrane permeation enhancers for macromolecular drugs. Laursen *et al.* studied the effects of three different concentrations (4, 8, and 16% w/w) of didecanoyl-L-alpha-phosphatidylcholine (DPPC) on the nasal absorption of human growth hormone (hGH) in normal subjects. They found that the absorption of hGH across the nasal mucosa accelerated with an increase in the concentration of DPPC [25].

3.2.4.4 Bile salt derivatives

Several bile salt derivatives such as sodium cholate, sodium deoxycholate, sodium glycocholate, sodium taurocholate, sodium taurodeoxycholate, and sodium glycodeoxycholate have been investigated for their efficiency in enhancing the nasal absorption of macromolecules [4]. Incorporation of 1% sodium glycocholate in an insulin solution significantly lowered the plasma glucose levels in healthy volunteers. Mixed micelles containing sodium glycocholate and linoleic acid significantly improved the absorption of insulin across the nasal mucosa.

3.2.4.5 Peptidase and protease inhibitors

The nasal mucosa expresses a significant amount of peptidase and protease enzymes, of which aminopeptidases are most abundant. Such enzymes are capable of metabolizing peptides and proteins such as enkephalins, insulin, and proinsulin. These enzymes, therefore, might play an important role in diminishing the overall absorption of these macromolecules. Various aminopeptidase inhibitors such as bacitracin, bestatin, and amstatin have been examined for their potential in improving the nasal absorption of macromolecules [26].

3.2.5 Nasal Formulations

Several drug delivery systems have been employed for the administration of proteins and peptide drugs through the nasal route. These include nasal drops, sprays, pledgets, powder insufflations, topical gels, emulsions, and ointments, each having their own advantages and disadvantages over the others [1]. Powder formulations are chemically stable, require smaller amounts of preservatives, and can efficiently administer large doses. Sprays generate better bioavailability than powder formulations [27]. Recently, bioadhesive polymers have been included in the nasal formulation since it decreases clearance of dosage form and thus improve nasal availability of peptide and protein drugs. Several delivery systems such as liposomes, microparticles, and nanoparticles are able to control the release of macromolecules at the site of administration. Further, it may be possible to deliver peptide and protein drugs specifically to their site of action by chemically modifying the surface of liposomes, microparticulate, and nanoparticulate systems. Several delivery systems were formulated to enhance the nasal absorption of insulin, one of the extensively studied proteins. Table 3.1 illustrates the various insulin formulations that were recently published.

Table 3.1 Nasal delivery of selected insulin formulations

Formulation	Animal model	Results	Year and reference
Solution	Mice	Intranasal administration of insulin was able to retard the progression of experimental streptozotocin (STZ)-induced diabetic peripheral neuropathy (DPN) in mice.	Francis *et al.* 2009 [28]
Solution	Male Sprague Dawley rats	Enhancer: L- or D- penetratin and L- or D-octaarginine. L- and D- penetratin increased the absorption of insulin significantly more than L- or D-octaarginine. Further, it was found that L-penetratin did not damage the integrity of the nasal epithelium.	Khafagy el *et al.* 2009 [29]
Multivesicular liposomes (MVLs)	Rats	*In vivo* studies in STZ-induced diabetic rats showed that the insulin-loaded chitosan MVLs, carbopol MVLs, and conventional liposomes decreased the plasma glucose level to a similar extent.	Jain *et al.* 2009 [30]
Hydroxypropyl methylcellulose (HPMC)-lyophilized inserts	Healthy male volunteers	2% HPMC-lyophilized inserts increased the nasal residence time of the formulation to 4–5 hours compared to the nasal residence time of 9.2 min of the conventional nasal spray formulation.	McInnes *et al.* 2007 [31]
Aminated gelatin microspheres (AGMS)	Rats	Insulin-loaded AGMS significantly improved the absorption of insulin in rats than when given as insulin suspensions.	Wang *et al.* 2006 [32]
Microparticles	Rats	Chitosan-TBA microparticles showed better bioavailability (6.9 ± 1.5%) of insulin as compared to unmodified chitosan microparticles (4.2 ± 1.8%).	Krauland *et al.* 2006 [33]
Nanocomplex (NC)	Rats	Polymer: Amine-modified poly(vinyl alcohol)-graft-poly(L-lactide) NC of insulin was able to decrease the plasma glucose level in fasted healthy rats by 20% after 50–80 min and by 30% in 75–95 min in STZ-induced diabetic rats.	Simon *et al.* 2005 [34]

Formulation	Animal model	Results	Year and reference
Chitosan microspheres	Rats	The intranasal administration of insulin microspheres formulated with 400 mg chitosan and 70 mg ascorbyl palmitate decreased the blood glucose level by 67% when compared to insulin given intravenously. The bioavailability of insulin was 44%.	Varshosaz *et al.* 2004 [35]

3.2.5.1 Nasal drops

Nasal drops are one of the most suitable and convenient systems for the delivery of peptide and protein drugs in the solution form. Drops instilled into the nasal cavity spread over a greater area than the nasal spray. Three drops are enough to completely cover the walls of the nasal cavity [29]. The main drawback associated with the nasal drop is inaccurate dose delivery. Further, rapid nasal drainage may also reduce the efficacy of the nasal drops. Molecules that can be readily absorbed are, therefore, more suitable for nasal drops. Drops were far more capable in depositing human serum albumin in nostrils than nasal sprays.

3.2.5.2 Nasal sprays

Nasal sprays are preferred dosage systems for protein and peptide drugs in a solution or suspension form. With the introduction of metered-dose nasal actuators and pumps, solution and suspension formulations were more accurately administered into the nasal cavity. These spray-delivering devices are capable of administering doses ranging from 25 to 200 µL. For delivery of suspensions, the selection and assembly of spray devices depend on the size and morphology of the drug particle and the viscosity of the formulation. Doses administered using nasal sprays are mostly delivered in the nonciliated regions and thus exposed to relatively slower nasal clearance than nasal drops. Proteins and peptide drugs that are absorbed slowly require a longer residence time and may, therefore, be formulated as nasal sprays. Desmopressin when delivered using metered-dose nasal sprays showed higher bioavailability than when delivered as nasal drops. Metered-dose nasal sprays are now available in the market for the systemic delivery of various peptides such as desmopressin, oxytocin, calcitonin, and buserelin owing to their ability to provide accurate dose delivery and high systemic drug availability [36]. The Food and Drug Administration (FDA) has approved salmon calcitonin (sCT) nasal spray (Miacalcin®, Novartis) for the treatment of postmenopausal osteoporosis and Paget's disease [26].

3.2.5.3 Nasal powders

Nasal powder dosage form is particularly useful for those proteins and peptide drugs where stable solution or suspension formulation is difficult to formulate. Powder formulations impart advantages such as better chemical stability and no need for preservatives. Factors affecting absorption from nasal powders include solubility, particle size, and aerodynamic properties of the active drug and/or excipients. Moreover, the active drug and/or excipients have the potential to cause nasal mucosal irritation and the grittiness of the powder formulation can add to the two main disadvantages of powder formulations — the possibility of nasal mucosal irritation and difficulties in delivering accurate doses to the target site [37]. Specialized and highly efficient delivery systems are now available that can accurately deliver metered doses of powder formulations into the nasal cavity. Nagai *et al.* studied nasal absorption of insulin in dogs and rabbits by administering insulin in adhesive powder and liquid dosage form. This study reported that insulin was much better absorbed with less mucosal irritation in powdered dosage forms than in the liquid form [38, 39]. The powdered dosage form of elcatonin-containing calcium carbonate prolonged the mean residence time of elcatonin in the nasal cavity, further increasing its nasal bioavailability [37]. The powder formulation of glucagon was prepared and studied by Teshima *et al.* in order to determine its efficacy for improving metabolic status in fatty liver syndrome patients with pancreatectomy. The results were compared with glucagon spray solutions. The spray solution produced better glucagon absorption; however, it was unstable and caused irritation to the nasal mucosa. On the other hand, the powder formulation was stable and did not cause any irritation, suggesting its usefulness in protein and peptide drug delivery [40].

3.2.5.4 Nasal ointments and emulsions

Agents with very low solubility and residence time in the nasal cavity can be administered in the form of nasal ointments or emulsions. These dosage forms, however, failed to gain attention because of their main disadvantages such as inaccurate dose delivery and patient noncompliance. Kararli *et al.* studied the potential of an emulsion formulation as an enhancer of the absorption of a renin inhibitor, *O*-(*N*-morpholino-carbonyl-3-L-phenylaspartyl-L-leucinamide of (2S, 3R, 4S)-2-amino-1-cyclohexyl-3, 4-dihydroxy-6-methylheptane(l), across the nasal mucosa in rats. This study reported that absolute bioavailability of this modified tripeptide was increased from 3–6% (obtained when the inhibitor was given in a PEG 400 solution) to 15–27% when administered as an oleic acid/mono-olein emulsion formulation [41].

3.2.5.5 Nasal gels

A nasal gel is a highly viscous jelly-like material with retarded flow at steady state. The benefits of a highly viscous dosage form for nasal drug delivery include prolonged residence time, reduction in postnasal drip, reduced leakage of the formulation in the anterior region of the nasal cavity, lower probability of taste development due to minimal swallowing, and reduced irritation to the nasal mucosa due to the addition of soothing excipients in the formulation [1]. For the past few decades, bioadhesive polymer gels have been extensively investigated for improving the absorption of proteins and peptides through the nasal mucosa. As the name suggests, bioadhesive polymer gels are made up of polymers with a unique property of forming a gel upon contact with the nasal mucosa. These polymers form a gel at the site of administration due to either an increase in temperature and ionic strength or interaction with calcium ions [7]. Several studies have been carried out in rats to examine the potential of bioadhesive polymer gels in reducing mucociliary clearance. Results from these studies clearly suggest that their bioadhesive formulations were able to lower mucociliary clearance and thus prolong nasal residence time [27]. Pluronic F127, a well-known thermogelling polymer, is liquid at 25°C whereas it forms a clear viscous gel at body temperature, 37°C. Pluronic F127 when applied on the nasal mucosa, reduces nasal clearance of fluorescent microspheres from the nasal cavity. Recently, a chitosan bioadhesive gel was prepared and studied as a delivery system for insulin through the nasal route. After intranasal administration of the gel in diabetic rats, there was an increase in the absorption of insulin. Furthermore, the release of insulin from the gel followed zero order kinetics. This result clearly suggests that bioadhesive gels may be used for controlling the delivery of insulin and other macromolecules [42].

3.2.5.6 Liposomes

In recent years, liposomes have been widely studied as an intranasal drug delivery system due to their advantages such as controlled drug release, targeted drug delivery by surface modification, high entrapment efficiency for small as well as large molecules with different hydrophobicity and pK_a values, reduced drug dosage, and minimal side effects. Drug loading and leakage of liposomes encapsulating desmopressin and their permeation across the nasal mucosa were investigated. Results from these studies revealed that the permeability of desmopressin across the nasal mucosa was enhanced with the insertion of positively charged liposomes followed by negatively charged liposomes and solutions [43]. The effect on the bioavailability of calcitonin from an intranasal administration of liposomes was examined, and results are compared with a calcitonin solution. Positively charged liposomes exhibit markedly increased absorption of calcitonin through the nasal mucosa as compared to negatively charged liposomes and the calcitonin solution [34]. Recently, ultraflexible liposomes as a nasal delivery system for sCT were evaluated by Chen *et al.* in rats. The hypocalcemic effects from both positively and negatively charged liposomes were much higher than from regular liposomes and sCT solution. Intranasal administration of sCT-loaded ultraflexible liposomes can decrease the serum calcium levels to the same extent as the subcutaneous administered for the first two hours [44]. Further, the application of surface-modified liposomes with mucosal adjuvants has been evaluated to determine their ability in enhancing nasal mucosal immunity. The potential of ovalbumin, uncoated liposomes, and liposomes coated with Man3-DPPE (a neoglycolipid containing mannotriose and dipalmitoylphosphatidylethanolamine) to induce mucosal response was examined after their intranasal administration in BALB/c mice. OVA-specific IgG and IgA antibody levels in serum obtained with Man3-DPPE-coated liposomes was much higher compared relative to uncoated liposomes and ovalbumin alone [45]. This observation clearly indicates that surface modification of liposomes with neoglycolipid can be a useful technique for the efficient delivery of macromolecules (33). Iwanaga *et al.* examined the potential of using liposomes for local drug delivery by exposing rat nasal mucosa to 5(6)-carboxyfluorescein (CF) in phosphate-buffered saline (PBS) and CF loaded in DPPC-liposomes. DPPC-liposomes caused 20- to 28-fold greater retention of CF on the surface of the nasal mucosa relative to the PBS solution [46].

3.2.5.7 Nanoparticles

Nanoparticles are polymeric particles of size ranging from 10 to 1,000 nm. Nanoparticulate delivery systems are prepared by biomaterials such as mucoadhesive natural or artificial polymers in order to prolong nasal residence time. Proteins and peptide drugs may be encapsulated inside the core and/or adsorbed or chemically attached to the surface. These particles constitute a highly efficient drug delivery system and provide numerous advantages such as controlled drug release, tissue- or cell-specific targeting via surface modification, enhanced enzymatic stability of proteins and peptide drugs in biological environment, improved drug loading, and interaction with the biological membranes. *In vitro* analysis has shown that uptake of the nanoparticles by the nasal tissue is highly dependent on size and surface characteristics. *In vivo* studies have shown that these particles are actively taken up by M cells and are further transported via the lymphatic tissue to systemic circulation [1]. For intranasal administration of proteins and peptide drugs, various polymeric materials such as degradable starch, dextran,

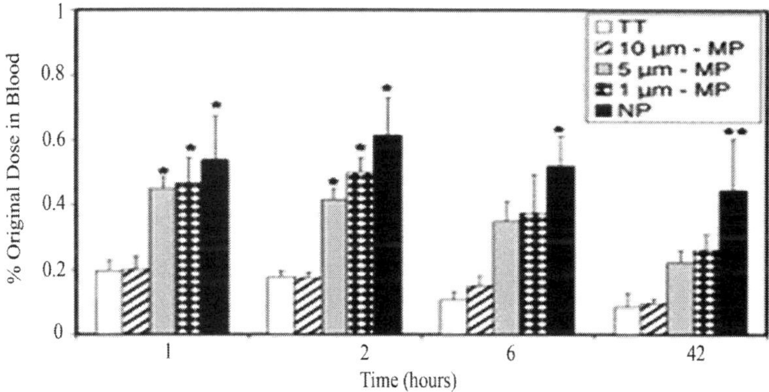

Figure 3.5 ^{125}I–TT blood levels following the nasal administration of free ^{125}I–TT or ^{125}I–TT encapsulated in PLA-PEG particles of different sizes (*significant differences ($p < 0.05$) with respect to TT and 10 µm-MP; (** significant differences with respect to TT, 10 µm-MP and 5 µm-MP).

chitosan, microcrystalline cellulose (MCC), HPMC, polycyanoacrylate, poly(D,L-lactide), poly(L-lactic acid), poly (lactide-co-glycolide), sodium alginate, and gelatin polymers have been selected for the preparation of nanoparticles. Figure 3.5 illustrates the effect of different poly (lactide-co-glycolide)-polyethylene glycol (PLGA-PEG) particle sizes (200 nm and 1.5, 5, and 10 μm) on the nasal transport of a model-encapsulated protein tetanus toxoid (TT). This study concluded that PLGA-PEG particles with a smaller particle size (200 nm) has higher nasal bioavailability (F = 70–80%) [47].

An insulin-loaded starch nanoparticulate formulation prepared with 0.2 mL epichlorohydrin (emulsion method) and sodium glycocholate markedly lowered the plasma glucose level by about 70%. Hypoglycemic activity was seen until six hours after the intranasal administration of insulin-loaded starch nanoparticles in rats. A combination of mucoadhesive nanoparticles and permeation enhancers can significantly enhance insulin absorption across the nasal mucosa [48]. Chitosan is the most extensively utilized polymer for nanoparticles containing insulin designed for nasal delivery. It is a cationic polymer and can, therefore, effectively interact with anionic sialyl group on the cell surface. This interaction may further enhance nanoparticle uptake. Insulin-containing chitosan nanoparticles were developed in order to raise the systemic availability following nasal administration. Nanoparticles of size 300–400 nm with a polycationic surface and a loading efficiency of 55% were prepared by ionotropic gelation of chitosan with tripolyphosphate anions. Nasal absorption was much higher with insulin-loaded chitosan nanoparticles relative to insulin in a chitosan solution [49]. Furthermore, chemically modified chitosans were employed for nanoparticle preparation due to their ability to enhance insulin uptake across the nasal mucosa. Such effect was much more pronounced than unmodified chitosan nanoparticles. The application of thiolated chitosan (chitosan-*N*-acetyl-L-cysteine) nanoparticles as a nasal delivery system for the delivery of insulin has also been examined in rats. *In vitro* mucoadhesive properties of thiolated chitosan nanoparticles were observed to be greater than those of unmodified chitosan nanoparticles. After intranasal administration, insulin-loaded thiolated chitosan nanoparticles increased insulin absorption across the nasal mucosa over insulin-loaded unmodified chitosan nanoparticles and insulin solution [50]. Chitosan-reduced gold nanoparticles have also been designed for nasal delivery of insulin. It was found that in rats, such nanoparticles were able to reduce the blood glucose to a level much lower than insulin-loaded borohydride-reduced nanoparticles. Chitosan-reduced gold nanoparticles can also be a promising delivery system for the nasal delivery of proteins and peptide drugs [51]. Insulin-loaded PEG-g-chitosan nanoparticles of size 150–300 nm were administered in rabbits through the nasal route. Results from this study clearly demonstrate a severalfold rise in the absorption of insulin with insulin-PEG-g-chitosan nanoparticles relative to insulin-PEG-g-chitosan suspension and insulin solution [52]. A recent report suggests that polysaccharide nanoparticles are able to reduce membrane transepithelial resistance, resulting in improved nasal permeability. The intranasal administration of insulin-loaded polysaccharide nanoparticles caused a marked reduction in plasma glucose levels in conscious rabbits [53].

3.2.5.8 Microparticles

Microparticles have been extensively investigated as a delivery system for the nasal absorption delivery of protein and peptide drugs. These particles are often prepared from mucoadhesive polymers in order to prolong nasal residence time on the nasal mucosa. Such delivery systems can also open up epithelial tight junctions and prevent the enzymatic degradation of proteins and peptides in the nasal cavity. Various gelling materials such as starch, albumin, chitosan, and DEAE-dextran have been employed for the preparation of microparticles. Particles prepared from such materials can easily adhere to the nasal mucosa by absorbing water on to their surface. The resulting gel prolongs mean residence time of the particles, which, in turn, improves absorption. Degradable starch microspheres (DSM) are one of the most popular microparticulate systems due to their biodegradable, biocompatible, and thickening properties. The nasal absorption of macromolecules such as hGH, desmopressin, and insulin improved significantly by DSM [4, 54, 55]. Further, a morphological examination revealed that DSM did not cause any noticeable histopathological changes to the nasal mucosa of rabbits during eight weeks of administrations [56]. The effects of DSM on the mucociliary clearance and geometry of the nasal cavities upon chronic administration in humans was studied by Holmberg *et al.* [57]. Repeated exposure to DSM did not cause any apparent changes in the mucociliary clearance or in the geometry of the nasal cavities

[29]. The safety of DSM was evaluated by administering particles in the nasal cavities of healthy human volunteers for a period of one week. Results have clearly shown that DSM is safe and well tolerated [4]. Hyaluronic acid microspheres of insulin were also prepared and studied for determining their potential in increasing the nasal absorption of insulin. Insulin bioavailability after their intranasal administration was comparable to a subcutaneously administered dose in the sheep. Starch and hyaluronic acid microspheres of insulin had different properties, but their efficiency in improving nasal absorption was comparable [58]. Illum *et al.* found that microspheres prepared from DEAE-dextran exhibit longer residence time in the nasal cavity. Nasal clearance of these microspheres is extremely slow. About 60% of the administered dose remained at the site of application after three hours (17). However, these microspheres were not effective in improving insulin absorption in rats (16).

3.2.6 Nasal Delivery of Vaccines

Microorganisms causing measles, pertussis, meningitis, and influenza can easily infect through the nasal mucosa. Most of the vaccines available for these microorganisms are in the form of injections, which may need proper storage conditions and may elicit adverse reactions. The nasal delivery of vaccines has gained significant attention because of its advantages such as ease of administration, a large mucosal surface area, and a high degree of vascularization. Further, needleless administrations, evasion of first-pass metabolism, ease and capability of mass vaccination, and faster onsets of sharp immune responses are the other benefits of nasal vaccination. Dose inaccuracy, nasal clearance, and nose-to-brain transport are some of the concerns expressed for nasal vaccine delivery [27, 59].

Multiple nasal vaccine systems have been studied with live or attenuated whole cells, proteins, polysaccharides, cholera toxin, attenuated virus, nanoparticles, microspheres, liposomes, and chitosan [59, 60]. For the nasal application of vaccines, particulate systems have been the most promising delivery system as these systems have the potential to achieve both mucosal and systemic immune responses. Particulate systems for the nasal delivery of vaccines have to conform to certain physicochemical parameters, that is, adequate size, surface charge, and hydrophobicity. These factors may largely determine their efficacy in inducing a high level of antibody response. TT-loaded nanoparticles (100 nm and 500 nm) induced higher serum IgG and IgA responses than 1,500 nm nanoparticles in female Balb/c mice [61]. Chitosan-antigen vaccines can cause a significant serum IgG response and generate greater secretory IgA levels in animals with influenza, pertussis, and diphtheria than injectable vaccines [62]. The nasal chitosan influenza vaccine is highly effective against flu viruses in humans [11]. Polymers that are highly hydrophobic in nature are the ideal materials for the nasal delivery of vaccines. An *in vitro* uptake study across Caco-2 cell line suggested that uptake of nanoparticles was enhanced with an increase in the hydrophobicity of polymers (poly-ε-caprolacton>co-polymer>PLGA-PCL> PLGA nanoparticles). After intranasal administration in BALB/c female mice, poly-ε-caprolacton nanoparticles encapsulating diphtheria toxoid induced maximum serum IgG antibody responses [63]. Recently, chitosan nanoparticles have also been employed for the delivery of vaccines through the nasal route. TT-containing nanoparticles were prepared by complexation between carboxymethyl cellulose and *N*-trimethyl chitosan polymers. These nanoparticles had an average size of 283 ± 2.5 nm with a loading efficiency of 95%. After their intranasal administration into mice, TT-loaded TMC/MCC nanoparticles induced higher IgG levels than MCC or TMC nanoparticles and TT in a PBS solution. Thus, TMC/MCC complex nanoparticles may be a promising delivery system for the nasal delivery of vaccines [64]. FluMist®, a nasal flu vaccine spray, has been developed and marketed for immunization against influenza virus type A and B [24]. The live, attenuated intranasal vaccine (LAIV) for H1N1 influenza virus has received FDA approval in September 2009 [26].

3.2.7 Intranasal Gene Delivery

In the past few decades, genes have been extensively studied for their potential as therapeutic agents. Gene therapy has been proposed for a wide number of systemic and local diseases. However, only a few of them have been successful in clinical trials. Recombinant genes used for therapies are very large molecular weight polar molecules; therefore, their administration requires a promising delivery system. For gene therapy to be successful, a highly efficient strategy should be chosen for delivering therapeutics to their target. The use of nasal route for the delivery of genes has become an area of high interest for both academic and industrial research. Table 3.2 illustrates the various nasal gene formulations and their references. Various viral and nonviral vectors have been investigated for the delivery of genes. Liposomal and polymeric delivery systems appear to be promising in delivering genes through the nasal mucosa. Several biodegradable and biocompatible polymers such as chitosan, PLGA, poly-L-lysine (PLL) have been selected for preparing nanoparticles and microparticles for gene delivery. Further, the surface and molecular modification of these particulate delivery systems appears to improve polymer biocompatibility, *in vivo* stability, and cell specificity. However, these particulate systems have low transfection efficiency as compared to the viral vectors [65]. The intranasal administration of interleukin-12 genes with polyethylenimine vector has been studied to examine its effectiveness in treating osteosarcoma lung metastases [66]. Hyde *et al.* studied the effect of repeated doses of DNA/liposomal formulation on the nasal epithelium in patients with cystic fibrosis. Selected DNA expressed cystic fibrosis transmembrane conductance regulator (CFTR) cDNA and was complexed with DC-Chol/DOPE cationic liposomes. The results showed no sign of inflammation, toxicity, or any immune response after multiple intranasal administrations [67]. Chitosan DNA nanospheres carrying a mixture of plasmid DNAs (specifically encoding for RSV antigens) significantly reduced the viral antigen load in mice with acute respiratory syncytial virus (RSV) infection. Moreover, these nanospheres caused a marked induction of RSV-specific IgG antibodies, nasal IgA antibodies, cytotoxic T lymphocytes, and interferon-gamma production in various regions of lungs and splenocytes [68]. Chitosan nanoparticles loaded with plasmid DNA were also prepared and tested for nasal mucosal immunization against hepatitis B. These nanoparticles successfully induced both systemic as well as mucosal immune responses after nasal administration [69]. Furthermore, chemically modified chitosans like thiolated chitosans were examined for gene delivery due to their excellent mucoadhesive and cell-penetrating properties. Thiolated chitosan (33 kDa) nanocomplexes with plasmid DNA (specifically encodes for green fluorescent protein) of size 70 to 120 nm resulted in a high expression of green fluorescent protein in HEK293, MDCK, and HepG-2 cells relative to unmodified chitosan. *In vitro* release studies clearly demonstrated a slow release of the cross-linked DNA, while *in vivo* data in Balb/c mice showed significant gene expression for 14 days [70].

Table 3.2 Selected intranasal gene delivery formulations

Gene	Formulation	Gene	Year and reference
CpG DNA	Cationic liposomes	Mice	Zhou *et al.* 2010 [71]
Plasmid DNA	PEG-grafted chitosans (CS-g-PEG)	Balb/c mice	Csaba *et al.* 2009 [72]
Anti-VEGF intraceptor (Flt23k) plasmid	Transferrin- or deslorelin-modified PLGA nanoparticles	Bovine	Sundaram *et al.* 2009 [73]
Murine T-bet gene (AAV-T-bet)	Solution	Mice	Wang *et al.* 2008 [74]
Luciferin	Solution	Mice	Buckley *et al.* 2008 [75]
Plasmid pRc/CMV-HBs(S)	Glycol chitosan–modified liposomes	Mice	Khatri *et al.* 2008 [69]
Leishmanial LACK gene (pCIneo-LACK)	Solution	Mice	Gomes *et al.* 2007 [76]
Plasmid DNA	Cationic emulsion	Mice	Kim *et al.* 2005 [77]
Rat insulin gene	Liposomes	Mice	Tanaka *et al.* 2004 [78]

Abbreviation: VEGF, vascular endothelial growth factor.

3.3 Pulmonary Drug Delivery

Extensive research efforts have been directed toward various noninvasive approaches to the systemic delivery of therapeutic proteins and macromolecules. Lungs are one of the most researched portals for noninvasive delivery. Pulmonary administration offers several advantages such as a large absorptive (100 m^2) surface area, the entire blood supply of the heart, a very thin absorptive membrane 0.1–0.2 μm, paucity of enzymes, extensive vasculature, and the short distance of air-blood exchange passage. Many peptides and proteins are absorbed readily through the alveolar cells in to the systemic circulation without the need for any permeation enhancer. In general, permeation enhancers possess surfactant activity that can damage the mucosal epithelial cells. For this reason, pulmonary administration has been the preferred route for macromolecular drugs.

The upper pulmonary region contains trachea, bronchial, and bronchioles, and the total surface area (100 m^2) is lined with a thick ciliated mucus-covered cell layer, making it nearly impermeable for macromolecules. On the contrary, the bronchial tree ends with large numbers of very small sacs known as alveoli where gas exchange takes place. These alveolar sacs are made of a thin monolayer, which can be easily traversed by macromolecules. In order to target these terminal alveolar cells, the delivery system must deposit its cargo to the lower part of the pulmonary tree. To reach that pulmonary segment, the deposited particles must be within an aerodynamic diameter range of 1–5 μm. The pulmonary delivery system may consist of a liquid spray with the drug dissolved in a solvent and/or propellant or one in which the drug can be delivered as solid particles through a dry powder inhaler. The dissolved drug molecules from the liquid spray can rapidly permeate the alveolar cells and be absorbed rapidly to elicit a rapid pharmacological response. However, solid drug particles must undergo rapid dissolution to be absorbed rapidly. Poorly dissolving solid particles may also undergo transcytosis into the lymphatic system and systemic circulation.

3.3.1 Anatomy of the Lungs

Airways originate from the trachea and bifurcate to form the main bronchi. These main bronchi divide to form smaller bronchi, which lead to individual lung lobes — three lobes on the right side and two on the left side. Inside each lobe, the bronchi undergo progressive divisions to terminal bronchioles, respiratory bronchioles, alveolar ducts, and finally to alveolar sacs. Airways are divided into conducting and respiratory segments (Fig. 3.6). The airway epithelium has a continuous sheet of cell types lining the luminal surface, which separates the external environment from the internal environment. Epithelial cells are connected by intercellular junctions that restrict the entry of foreign substances. Conducted airways are lined with ciliated columnar cells. Epithelial cells are covered by mucus, a viscous fluid containing glycoproteins secreted by goblet cells and mucus glands. However, the mucus layer is not secreted at the lungs periphery as goblet cells are not present in distal airways. Alveolar type 1 cells are the principal cell type that lines the alveoli. Gas exchanges take place through these cells, that is, oxygen and carbon dioxide exchange in capillaries. Alveoli type 2 cells are progenitor cells that regenerate alveolar type 1 cells. These cuboidal cells express surface microvilli and secrete surfactants lining the alveolar epithelium. Surfactants comprised a mixture of carbohydrates, proteins, and lipids.

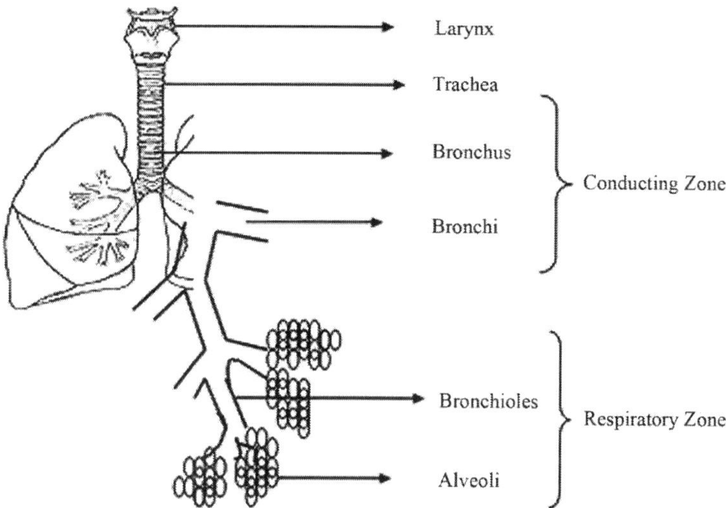

Figure 3.6 Anatomy of lungs.

Smooth muscle cells are separated from the epithelium by the lamina propria, a region of connective tissue containing nerves and blood vessels. The contraction of this smooth muscle is the primary cause of obstructive airway diseases. Goblet cells that secrete mucus into the airway lumen are located in the submucosa. Each mucus gland consists of four regions: the ciliated duct, the collecting duct, mucus tubules, and secretory tubules. The ciliated duct ends in the airway lumen and is lined by ciliated epithelial cells. Columnar cells line collecting ducts whereas mucus cells line mucus tubules. Several secretory tubules feed into the collecting duct. Airways are innervated by afferent and efferent nerves that serve sensory and efferent functions in the central nervous system involved in the regulation of airway function. Alveolar macrophages present in the interstitium and luminal surface of the alveoli engulf foreign substances, particles, and toxins. The activation of alveolar macrophages and mast cells results in the release of inflammatory mediators, enzymes, and cytokines. The pulmonary artery carrying the oxygenated blood divides into intrapulmonary arteries and further into pulmonary capillaries. These capillaries unite to form small veins, which drain into the pulmonary vein, which delivers blood from lungs to the heart.

3.3.2 Pulmonary Absorption

Lungs have a large surface area, with approximately 80–100 m^2 available for drug absorption. In addition, lungs possess a thin alveolar epithelium (100–200 nm) and are highly vascularized, with an extensive capillary network. Lung vessels also receive entire cardiac output of venous blood. Three main mechanisms of pulmonary drug transport have been proposed: (1) paracellular transport through membrane pores, (2) vesicular transport, and (3) transcellular transport through intercellular junctions. Small molecules can be absorbed either by diffusion or by carrier-mediated transport. Small molecules (<1 kDa) are rapidly absorbed to the blood stream through paracellular pathway. High-molecular-weight compounds may enter the blood stream via the transcellular pathway, or pinocytosis. Macromolecules with a molecular weight less than 40 kDa may be absorbed by paracellular transport whereas molecules larger than 40 kDa may be absorbed by transcytosis. Even though lungs represent rapidly perfused organs with high absorption efficiency, studies have shown that proteins and other macromolecules exhibit low bioavailability possibly due to the loss in the airway lumen and epithelium. Insulin (5.8 kDa) has an average 10% bioavailability relative to subcutaneous injection whereas a high molecular weight protein like IgG (150 kDa) delivered intratracheally has a significantly lower bioavailability (5%). The central airway epithelium expresses ciliated columnar cells with tight intercellular junctions. Therefore, the absorption primarily occurs through paracellular diffusion as well as carrier-mediated processes. The rate of diffusion across the central airway epithelium is directly proportional to the concentration gradient and lipid solubility and inversely proportional to the molecular size and ionization. A significant fraction of the inhaled drug may be lost due to binding with the constituents of the mucus and subsequent clearance by mucociliary apparatus. The alveolar epithelium has cuboidal epithelial cells lined with surfactant. Molecules deposited in the surfactant layer may diffuse laterally, thereby lowering the local drug concentrations. The absorption of compounds across the alveolar epithelium depends upon the molecular size and lipophilicity or hydrophilicity. Low-molecular-weight compounds dissolve rapidly and diffuse readily through the alveolar epithelium. The pore diameter in the epithelium and endothelium are 0.6–1.5 nm and 2–13 nm respectively. The absorption rate also depends upon the solubility of inhaled molecules at the absorption site. Wettability, agglomeration, ionization, crystallanity, and polymorphism of particles that alter drug solubility can also modify the rate and extent of absorption. Alveolar macrophages and proteases can also alter the rate and extent of absorption.

3.3.3 Barriers in Pulmonary Drug Delivery

Higher bioavailability across the alveolar epithelium renders the pulmonary route as one of the preferred alternative noninvasive routes for systemic or local therapy. However, there are several barriers that restrict drug absorption into the systemic circulation. Since lungs are continuously exposed to airborne pathogens, the tissues possess several complex defensive mechanisms such as mucociliary clearance, coughing, and alveolar clearance. Mucociliary clearance is the main clearance mechanism of conducting airways, which clears inhaled particles by coordinated interaction between the thick mucus (1–10 μm) secreted by goblet cells and beating of the airway cilia (5–6 μm). Inhaled particles deposited on the mucus are transported to the pharynx in the upward direction by the airway cilia. Normal mucociliary transport occurs at a rate of 5.5 mm/min at the trachea and 2.4 mm/min in the major bronchi [79]. The alveolar epithelium, basement membrane, pulmonary enzymes, secretory immunoglobulin A, dendritic cells, and macrophages constitute the alveolar clearance in this region [80]. Alveolar macrophages primarily consist of lung phagocytes in airways, interstitial matrix, and alveolar regions. Inhaled particles are rapidly cleared from the alveolar region either by the mucociliatory apparatus, to the tracheobronchial lymph, and/or by enzymatic degradation. This clearance from the alveolar region is slower than mucociliary clearance and depends upon the particle size. A thin layer of surfactant (0.1–0.2 μm) covering the alveolar epithelium acts as a physical barrier to drug absorption. Inhaled particles bind to this surfactant and aggregate, resulting in enhanced clearance. In addition, proteins and peptides are also degraded by proteases, peroxidases, and inflammatory mediators secreted by macrophages as a part of the host defense mechanism.

3.3.4 Formulations

Macromolecules can be encapsulated in a variety of nanostructures such as micelles, liposomes, nanoparticles, and microemulsions, which are then accurately delivered by metered-dose inhalers, dry powder inhalers, and nebulizers. Absorption enhancers, protease inhibitors, cyclodextrins, surfactants, and tight junction modulators are utilized to improve the pulmonary absorption and stability. Yamamoto *et al.* studied the effects of absorption enhancers such as sodium glycocholate, sodium caprate, linoleic acid-surfactant mixed micelles, *N*-lauryl-β-D-maltopyranoside, and ethylenediaminetetraacetic acid (EDTA) as well as protease inhibitors such as aprotinin, bacitracin, and soyabean trypsin inhibitors on the pulmonary absorption of insulin and eel-calcitonin [81]. This study reported the improvement in pulmonary absorption of insulin measured by hypoglycemic response after peptide administration. Table 3.3 illustrates the various insulin formulations and their effects on pulmonary absorption and plasma glucose levels.

3.3.4.1 Micelles

The size and shape of aerosol particles play an important role in the pulmonary deposition and permeability. Micelles are spherical nanostructured particles (10–100 nm) formed from the self-assembly of amphipathic molecules in water. Hydrophobic molecules align at the tail regions whereas hydrophilic molecules form the central core. Micelles can be surface functionalized to modify targeting, release, and clearance. Polymeric micelles formed by the self-assembly of ampiphilic block copolymers improve the solubility of hydrophobic molecules as well as enhance their tumor accumulation and prolong circulation time *in vivo* [82]. Paclitaxel-loaded polymeric micelles (Genexol-PM) combined with cisplatin are currently in phase II clinical trials for the treatment of advanced non-small-cell lung cancer (NSCLC). Polymeric micelles surface functionalized with lung cancer targeting peptides (LCP) that bind to integrins have been investigated for targeted delivery to lung cancer [83]. Bhattarai *et al.* developed an aerosol system containing a novel ampiphilic triblock copolymer [poly(*p*-dioxanone-co-L-lactide)-block-poly(ethylene glycol)] for the topical delivery of tumor suppressor gene phosphatase and tensin homolog (PTEN) to lungs of C57BL/6 mice. This aerosol system significantly reduced tumor size and lung metastasis relative to control mice. Recently, rhodamine-loaded immunomicelles attached with antitumor antibody 2C5 have shown enhanced accumulation in Lewis lung cancer cells [84].

3.3.4.2 Liposomes

Liposomes have many potential advantages such as protection from clearance, sustained pulmonary release, targeting of encapsulated macromolecules, and subsequent internalization. Liposomal aerosols have been utilized to enhance drug retention time and control drug release in lungs with reduced toxicity. Lungs secrete a complex mixture of surfactants, out of which 75% is dipalmitoylphosphatidylcholine. The mechanism of surfactant uptake and clearance play an important role in the internalization of liposomes. Previous studies utilized gamma scintigraphy to study deposition and clearance of radiolabeled liposomes in the human respiratory tract. These studies have shown that liposomes can be deposited efficiently in lungs and can remain intact for a prolonged period of time. Also, the site of deposition depends upon the size of aerosol particles produced by the nebulizer. Inhaled liposomal aerosol formulations improved antioxidant activity of CuZnSOD (3500U) by about 125%. Moreover, the entrapment efficiency and activity of copper-zinc superoxide dismutase (CuZnSOD) are higher with anionic liposomes over cationic and neutral liposomes. Briscoe *et al.* encapsulated superoxide dismutase in pH-sensitive liposomes made up of 1, 2-dioleoyl-snglycero-3-phosphoethanolamine (DOPE) and 1-oleoyl-2-oleoyl-snglycero-3-succinate (DOSG). Upon internalization, pH-sensitive liposomes enter the cell and the liposomal contents are released in the acidic environment present in the endosome. Proteins such as catalase, superoxide dismutase, and glutathione have been encapsulated in liposomes and administered intratracheally by using nebulizers. However, due to long-term instability problems, proteins need to be encapsulated in dry powder liposomes. Lu *et al.* selected β-glucuronidase as a model protein to evaluate the effect of lyophilization, micronization, and aerosolization. This study demonstrated the release of 73% of the dose [liposome and mannitol (1:19)] from the dry powder inhaler [85]. Bi *et al.* studied the hypoglycemic effect of insulin-loaded liposomes prepared by spray-freeze drying in diabetic rats. *In vivo* results demonstrated a 38.33% relative bioavailability increase with the intratracheal administration of 8 IU/kg dose [86]. Other studies reported that octarginine peptides and lectins can enhance bioadhesion and airway cell internalization [87, 88]. The delivery of negatively charged plasmid DNA, small interfering RNA (siRNA), and antisense oligonucleotides requires cationic liposomes. Adenoviral vectors carrying p53 gene have already been evaluated in clinical trials. However, these have limited application due to their immunogenicity, high toxicity, and limited methods of administration [89]. Also, plasmid DNA needs to be protected from shear forces of nebulization. Cationic liposomes have been used in gene therapy for acute respiratory distress syndrome, lung transplantation, lung cancer, cystic fibrosis, pulmonary hypertension, alpha-1 trypsin deficiency, and asthma [90]. Cationic liposomes have been proven to be effective in delivering aerosol genes due to interaction with negatively charged DNA. Moreover, the addition of fusogenic lipids such as cholesterol and DOPE can facilitate endosomal release at low pH. Recently, Garbuzenko *et al.* demonstrated enhanced peak concentrations and tissue retention in lungs with intratracheal delivery of neutral and cationic liposomes containing doxorubicin, antisense oligonucleotides (ASO), and siRNA. This study utilized neutral PEGylated liposomes and 1,2-dioleoyl-3-trimethylammonium-propane (DOTAP) cationic liposomes to demonstrate intracellular internalization in human non-small-cell lung carcinoma cells (A549) and in nude mice with an orthotopic model of lung cancer [91]. Spray-dried powders have been proven to be effective in enhancing and preventing drug loss. Moreover, this system can be targeted to pulmonary regions without any loss of biological activity. Further, various thermoprotective additives such as carbohydrates and amino acids are added to cationic lipid complexes to obtain stable and viable formulations with minimal damage to pDNA [92].

3.3.4.3 Microparticles

Particulate delivery systems, particularly microparticles, and nanoparticles have been extensively investigated for the pulmonary delivery of macromolecules. Nanoparticles are polymeric particles in the nanometer size range (10–1,000 nm) whereas microparticle sizes are in the micrometer range (1–1,000 μm). Microparticulate systems include microspheres (uniform spherical particles with a polymeric matrix) and microcapsules (heterostructures with an oily core containing an active material surrounded by thin polymeric membrane). These carrier systems are ideal for the delivery of macromolecules with a short half-life *in vivo* and suffer from poor bioavailability at the target site. In addition, such systems can encapsulate both hydrophilic and hydrophobic molecules, sustain release, and improve stability, biodistribution, and drug targeting. Due to a high surface/volume ratio, these particles can be surface functionalized with cell internalization peptides and antibodies as well as receptor-targeted ligands to modulate biodistribution *in vivo* and subsequent target cell internalization. For pulmonary delivery, particle sizes of 1–5 μm render them ideal for phagocytosis by macrophages and agglomeration due to van der Waals and electrostatic attractions. Large porous microparticles prepared by methods such as emulsion-solvent evaporation, spray-drying, emulsion-solvent diffusion, phase separation, and vapor condensation techniques usually possess large geometric diameters but a low density and small aerodynamic diameters [93].

3.3.4.4 Nanoparticles

Nanoparticles are made of polymers [natural polymers (gelatin, albumin); synthetic polymers (polylactides, polycyanoacrylates)] or lipids. Therapeutic agents can be bound either on the surface of the particles or encapsulated inside the polymers. Inhaled nanoparticles provide several advantages over other carrier systems due to their high stability and carrier capacity, encapsulation of both hydrophilic and hydrophobic substances, and feasibility of targeting by modulating the surface and controlled release. Nanoparticles consist of PLGA polymers, which are biodegradable and biocompatible. Nanoparticle surface properties and size determine their airway deposition and evasion of physiological barriers such as mucus and mucosal layer. Noninvasive pulmonary delivery of nanoparticles can enhance and sustain local drug concentrations in the lung tissue as well as systemic circulation, which may reduce dosing frequency and improve patient compliance. Edward *et al.* have shown that large porous PLGA insulin nanoparticles with a diameter greater than 5 μm and mass density less than 0.4 gram per cubic centimeters are more therapeutically effective than small, porous particles. This study reported elevated systemic levels of insulin and suppressed systemic glucose levels over 96 hours with large porous nanoparticles whereas small nonporous nanoparticles produced hypoglycemic activity only for 4 hours. The relative insulin bioavailability is 87.5%, which is about seven times higher than small and nonporous nanoparticles [94, 95]. Large particles can be delivered more efficiently from an inhaler, resulting in higher respirable fractions. Also, such nanoparticles can avoid macrophage clearance. But, later studies have shown that for effective lung delivery, particle size and mean aerodynamic diameter should be less than 5 and 2 μm, respectively. Ultrafine nanoparticles less than 100 nm produced by micronization readily translocate the pulmonary epithelium into blood stream by avoiding macrophagal clearance. Therefore, a balance must be struck between the particle size, deposition, and toxicity profiles for systemic as well as local delivery. Several manufacturing methods such as spray drying, freeze drying, spray-freeze drying, supercritical fluid technology, double emulsion/solvent evaporation technology, antisolvent precipitation, and particle replication in nonwettable templates are some of the parameters utilized to optimize the deposition of nanoparticle formulations [96]. Spray drying is the predominant manufacturing method for the preparation of inhalable peptide particles. Inhaled nanoparticles are less favored to microparticles because of the propensity to be exhaled and the ability to escape from the lungs due to their smaller mass and median aerodynamic diameter. Also, these particles tend to aggregate during the manufacture of liquid and dry powder formulations. To solve this problem, Jeffrey *et al.* incorporated gelatin and polybutylcyanoacrylate nanoparticles into the matrix of carrier microparticles that can deaggregate upon dissolution in the aqueous environment of the lung. Lactose is usually recommended as an excipient. After spray-drying, these particles were found to be within the optimum range (3.0 ± 0.2 μm). Recently, Grenha *et al.* developed insulin-loaded chitosan/tripolyphospate nanoparticle formulations for pulmonary delivery. This study demonstrated an acceptable protein-loading capacity (65–80%) and a 75–80% insulin release within 15 min of aerosol administration [97]. Cationic polymers such as polyethylenimine have been used as a gold standard in nonviral gene delivery. Polyethylenimine (PEI) has a large charge density due to the presence of primary, secondary, and tertiary amines, which bind plasmid DNA compactly to the nanoparticles. In addition, their targeting ability can be enhanced by attaching several ligands such as antibodies, peptides, transferrin, lactoferrin, sugars, folic acid, and mannuronic acid. Recently, Elfinger *et al.* linked surface-adsorbed insulin peptides to PEI-pDNA nanoparticles to target insulin receptors present on the alveolar epithelial cells. Surface adsorption with insulin enhanced gene expression by 16-fold in alveolar epithelial cells but not in bronchial epithelial cells [98]. Zhang *et al.* prepared pulmonary surfactant nanoparticles to deliver insulin dry powder. A pulmonary surfactant is a complex mixture of phospholipids and proteins that reduces surface tension at the alveolar air-liquid interface. In this study, a pulmonary surfactant is used as an absorption enhancer. Upon inhalation of dry powder formulations, the relative pharmacological bioavailability for control and pulmonary surfactant nanoparticle formulation group is 19.97% and 32.57%, respectively. Animal studies have shown that insulin dry powder of pulmonary surfactant nanoparticles reduced the blood glucose levels significantly more than dry powder insulin and subcutaneous injections (Fig. 3.7).

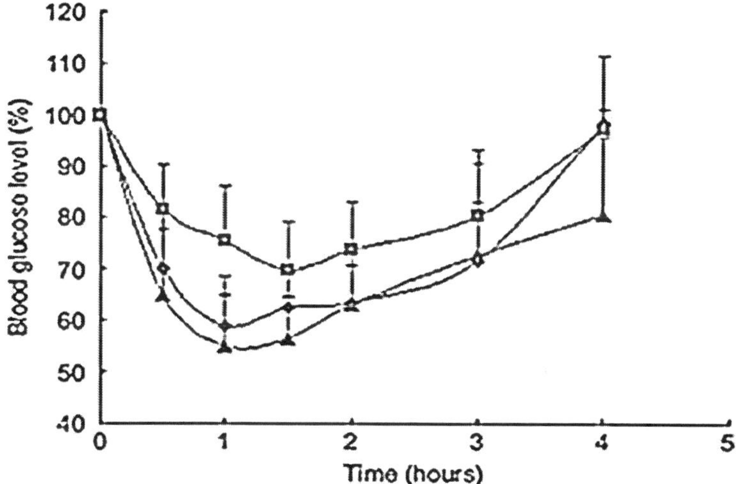

Figure 3.7 The blood glucose level-time curve of rats that received 4 U/kg SC free insulin (solid diamond), 14 U/kg dry powders of control group (solid rectangle), or 14 U/kg pulmonary surfactant group (solid triangle). Data is expressed as means (n = 6).

3.3.4.5 Microemulsions

Microemulsions have been studied for the pulmonary delivery of macromolecules. Microemulsions (10–200 nm) are thermodynamically stable dispersions of water and oil stabilized by a surfactant and a cosurfactant. In comparison to other nanocarrier systems, microemulsions offer several advantages such as higher drug encapsulation, stability, improved solubility, and ease of manufacture. Lecithin inverse microemulsions encapsulating water soluble compounds generate aerosols with sizes ranging from 2.7 to 3.1 μm and fine particle fraction values of 50–70% [99]. Poly(ethyl cyanoacrylate) nanocapsules prepared from w/o microemulsions are prepared to determine the release rate of insulin. Nanocapsules prepared by this method generate a mean particle size of 150.9 nm and encapsulation efficiency of 80% (73).

Table 3.3 Pulmonary delivery of selected insulin formulations reported in literature

Formulation	Animal model	Results	Year and reference
PLGA microcapsules	Rats	Fivefold longer residence time than subcutaneous injection of neutral protamine Hagedorn (NPH) insulin	Hamishehkar *et al.* 2010 [100]
Solid-lipid nanoparticles	Rats	Entrapment efficiency 69.47 ± 3.27%; prolonged hypoglycemic activity and relative pharmacological bioavailability 44.4%	Bi *et al.* 2009 [101]
Microcrystal powder	Male Sprague Dawley rats	Relative bioavailability of insulin microcrystals and insulin solution –15% and 10% respectively when compared to subcutaneous injection	Lim *et al.* 2009 [102]
Pulmonary surfactant nanoparticles	Rats	Relative bioavailability of inhaled insulin dry powder for control group and surfactant group found to be 19.97% and 32.57%, respectively, when compared to subcutaneous injection	Zhang *et al.* 2009 [103]

Formulation	Animal model	Results	Year and reference
Liposomes	Rats	Liposomes with DPPC-enhanced pulmonary insulin delivery in rats	Chono *et al.* 2009 [104]
Cationic cell-penetrating peptides conjugated to insulin	Rats	Oligoarginine-tagged insulin-enhanced pulmonary absorption of insulin in diabetic rats and blood glucose levels more sustained over control	Patel *et al.* 2009 [105]
PLGA/ hydroxypropyl-beta-cyclodextrin	Diabetic rats and hypoglycamic rats	Significant reduction of blood glucose level for prolonged time on delivery of large porous particles of insulin through PLGA/ cyclodextrins	Ungaro *et al.* 2009 [106]
Dry powder liposomes	Diabetic rats	Relative pharmacological bioavailability found to be 38.38%	Bi *et al.* 2008 [86]
Solid-lipid nanoparticles	Rats	Relative bioavailability 22.33% more than that through subcutaneous injection of insulin	Liu *et al.* 2008 [107]
Liposomes	Mice	Plasma glucose levels effectively reduced with inhaled insulin-loaded liposomes	Park *et al.* 2007 [108]
Microcrystal suspension	Rats	Enhanced hypoglycemic activity of insulin microcrystals because of the addition of zinc as an adjuvant	Huang *et al.* 2007 [109]
Micronized dry powder	Humans	Relative bioavailability from 11.5 to 12.2% for four doses of insulin (60, 90, 120, and 150 U)	Rave *et al.* 2004 [110]

3.3.5 Inhalation Devices

Aerosol inhalation systems such as metered-dose inhalers, nebulizers, and dry powder inhalers have been widely employed for the treatment of obstructive airway diseases. Macromolecules such as proteins and genes have a narrow therapeutic dose and need to be delivered in a reproducible manner to the lung periphery at a dose between 2–20 mg [1]. These devices have several limitations regarding delivery of macromolecules, such as low system efficiency, low drug mass per puff, poor formulation stability, and poor reproducibility. The choice of device depends on the drug, the formulation, the site of action, and the pathophysiology. The optimum particle size depends upon device variables (valve design, metering volume, spray orifice diameter, actuator design) and formulation variables (propellant composition, drug concentration, cosolvent content, and surfactant contents). The aerosol formulation is specific to a compound and has to generate an optimum particle dose with a mass median aerodynamic diameter of 1–5 μm. Current aerosol delivery devices include nebulizers, metered dose, and dry powder inhalers. Table 3.4 summarizes the various differences between these three types of devices [111–113].

Table 3.4 Comparison of the various properties of three types of delivery devices

Device specification	Pressurized metered-dose inhalers	Nebulizers	Dry powder inhalers
Components	Is the most common device that delivers either particulate suspensions or solutions	Sprays a fine liquid mist of medications	Delivers dry powder formulations
	Formulation, propellants, metering valve, and spry actuator	Compressor, compressor air or oxygen, mask, and nebulizer	Powder formulation, a dose-measuring system, powder deagglomeration principle, and mouth piece
Propellants	Hydrofluoroalkanes propellants used to push medication out of the inhaler	No propellants; requires strong and fast inhalation	No propellants; requires strong and fast inhalation

Device specification	Pressurized metered-dose inhalers	Nebulizers	Dry powder inhalers
Breathing coordination	Requires significant coordination, which can be minimized with the use of spacers	Requires minimal coordination	Needs coordination between breathing and actuation
Cleaning and maintenance	Does not need maintenance	Needs periodical cleaning due to bacterial contamination	Does not need maintenance
Cost and portability	Low cost, small, and portable	Bulky, costly , time consuming, and dependent an external power source	Low cost, small, portable, and easy to use
Usage	For children under four years, chambers with face masks required	Used for all ages	Used for patients over six years
	Cannot be used in mechanically ventilated patients	Can be used in mechanically ventilated patients	Cannot be used in mechanically ventilated patients
Dosage	High dose-dose reproducibility	Single-dose capability	Poor dose reproducibility
	Multidose capability	High dose and dose modification possible	Single dose and multidose types depending on design
Deposition	High pharyngeal deposition without spacer	Lower pharyngeal deposition	Inefficient deposition in diseased states
	Protein denaturation due to propellants	Inefficient delivery and protein denaturation due to high shear stress	High pharyngeal deposition
Examples	Autohaler, Easibreathe, K-Haler, Xcelovent, Smartmist, Easidose, Breathe Coordinated inhaler, Spacehaler	Aerogen, Omron, ODEM, PARI, Respimat, Respironics, Beurer	Spinhaler, Rotahaler, Aerolizer, Handihaler, Diskhaler, Turbuhaler, Diskus

3.3.5.1 Pressurized metered-dose inhaler (pMDI)

A metered-dose inhaler is one of the most frequently recommended devices for administering inhaled anti inflammatory steroids, β-agonists, and anticholinergics. These devices use trichlorofluro-methane, dichlorodifluromethane, and dichlorotetrafluoroethane as propellants to deliver a metered dose of a drug either in suspension or solution. Pressurized metered-dose inhalers can provide local effects in conducting airways due to small size. Portability, convenience, and lower cost are the other added advantages. The suspension formulations contain micronized drug particles dispersed in combination of volatile propellants whereas solution formulations contain drug particles that freely dissolve in a volatile propellant or a combination of propellant and cosolvent. The aerosols are often released with high velocity, and to optimize the release and deposition, spacers are often used. Tube-, cone-, and pear-shaped spacers are selected to overcome the problems of poor coordination and oropharyngeal deposition [2]. However, these devices are not applicable to proteins, since such molecules are prone to degradation upon contact with propellants. These devices also suffer from other limitations, such as low lung disposition and inefficient hand-mouth coordination. Also, the propellants generate a high velocity cloud (>30 m/s) that causes massive deposition, which might interfere with the inspiration.

3.3.5.2 Nebulizers

A nebulizer is an inhaler device that sprays a fine liquid mist of medication. Nebulizers can be readily applied for preexisting macromolecular formulations that can be easily aerosolized. A large percentage of the aerosols delivered by a nebulizer is deposited in the lungs with minimal gastrointestinal absorption, which is the case in metered-dose inhalers and dry powder systems. Two types of nebulizer systems are usually indicated — the ultrasonic and air jet nebulizers. In an ultrasonic nebulizer, waves are formed in an ultrasonic chamber through electronic excitation of a ceramic crystal. These high-energy waves disrupt the liquid surface and generate a high-concentration aerosol at the solution surface. In case of air jet nebulizers, compressed air is passed through an orifice, creating a low-pressure area where the air jet exists. The liquid is then drawn from the nozzle to mix with the air jet to form droplets of mist. These nebulizers have been successfully applied for stable pharmaceutical compounds. However, aerosol particles generated by these devices are not appropriate for proteins and peptides as these molecules are denatured irreversibly. Such devices deliver a lower drug dose per unit time relative to dry systems. The devices need refrigeration due to reported stability problems and microbial growth. Nebulizers are widely utilized for the delivery of gene therapy formulations. DNase is the only protein approved for delivery by inhalation through jet milling.

3.3.5.3 Dry powder inhalers

Dry powder inhalers are usually indicated for the treatment of local and systemic diseases because these systems are propellant free with high patient compatibility and high dose-carrying capacity. Since these devices do not use CFC propellants, they are preferred due to formulation stability and ease of use. These devices may be breath actuated or driven by external energy. Conventional methods generate fine powder dispersions by crystallizing the powder followed by disaggregation of micronized particles by milling. Spray-drying, spray-freeze drying, solvent precipitation, and supercritical fluid technology have been explored to optimize the crystallization, particle size, shape, and distribution. The particle size of the powder dispersed should have an aerodynamic diameter of 0.5–3 μm for deposition into the alveolar region of the lung to produce a systemic effect and 3–5 μm for local action [114]. There are two types of dry powder inhalers — unit dose and multidose. In case of unit dose devices, the drug is mixed with the bulk carrier and filled in a gelatin capsule. Whenever the device is actuated, the gelatin capsule is pierced, resulting in inhalation of the dose by the patient. Multidose devices work as pMDI and are convenient due to their size and compact design. Exubera, developed by Nektar, was delivered through this device. Recent studies have shown that dry powder inhalers were able to deliver inhalable macromolecules with a median mass aerodynamic diameter (MMAD) of 0.5–4.0 μm and 60% lung deposition. However, delivery of macromolecules through this device encounters stability problems due to micronization and disaggregation. These devices are highly complex, and the performance depends upon the inhaler performance, formulation, and air flow resistance. For effective deposition, the inspiration flow rate has to be 30 L/min, which might not be achieved by patients with chronic lung disorders.

3.3.6 Factors Affecting Pulmonary Deposition

An aerosol contains solid or liquid particles dispersed in a gaseous medium. About 80–95% of the inhaled dose reaches the gastrointestinal tract. Irrespective of the delivery method and device, a specific minimum dose (mass of drug deposited in the lung) is required to produce a clinical effect. There are two basic factors that affect the deposition of an aerosol — aerosol properties and patient factors. Internal impaction, sedimentation, diffusion, interception, and electrostatic precipitation are the five basic physical mechanisms that cause the deposition of aerosols. A combination of these mechanisms plays a vital role in the deposition of inhaled particles in different regions. A patient's ability to use and coordinate the inhaler device in a proper manner as with metered-dose inhalers and breath-driven powder generators plays an important role in the aerosol delivery. Also, anatomy and spatial patterns of airways affect the clearance of inhaled materials. Only a small fraction of the inhaled dose is retained in the alveolar region after 24 hours of deposition. The inhaled dose is cleared from the tracheobronchial tree primarily due to mucociliary clearance mechanism. Size, density, hygroscopicity, electrical charge, chemical properties, and aerodynamic diameters of inhaled particles need to be considered while designing the delivery of inhaled aerosols. Particles with aerodynamic diameters greater than 15 μm are usually deposited in the head or extrathoracic region. Particles between 5–10 μm get deposited mostly (20% deposition) in the tracheobronchial region. Maximum alveolar deposition (60% deposition) can be seen with particles between 1–3 μm. Submicronic particles lesser than 0.5 μm have a negligible probability of getting deposited in the head and bronchial regions.

3.3.7 Vaccines

More than 100 vaccines have been approved by the FDA for the treatment of respiratory infections. Pulmonary vaccinations adequately immunize patients against respiratory viral and bacterial infections. Vaccines offer defense against pathogens that enter via the mucous membrane in lungs. Aerosol vaccines do away with the need for needle vaccine, prevent the spread of HIV caused by infected needles, and are more patient compliant [115]. Aerosol vaccines are proven to be more effective than intranasal vaccines and parenteral injections since lungs contain a highly responsive immune system. The interstitial and peripheral tissues are abundant in alveolar macrophages and dendritic cells and are responsible for eliciting both innate and adapted immunity. The airway epithelium plays a crucial role in the host defense as it displays a mucociliary clearance mechanism. In addition, it secretes antimicrobial agents, cytokines, and chemokines to prevent the colonization of microorganisms. The mouse model is an ideal choice to study vaccine delivery because of its well-characterized immune system and the ability to create strains with the desired immune system, such as gene knockout mice. Measles vaccination via pulmonary aerosol delivery was the first successful clinical case with a high rate of prevention [116]. Dry powder aerosol vaccination of recombinant influenza virus demonstrated 89% survival in mice [117]. Nebulizers have been selected to deliver live, attenuated pathogens such as tularemia, measles, Bacillus Calmette-Guerin, and rubella. However, there were reports of loss of viral potency due to degradation caused by jet nebulization [118]. Vaccine adjuvants have been added to stimulate immune response. These agents can be employed to enhance the immunogenicity of antigens, prolong the immune response, and selectively modulate to a desired immune response, that is, major histocompatibility complex I/II, cytotoxic, and helper T lymphocytes. Aluminum hydroxide has been selected as an adjuvant in the intratracheal administration of ricin toxoid in rats. Particulate delivery systems such as liposomes and biodegradable microspheres have been employed to target mucosal-associated lymphoid tissues (MALT) so that antigens can be actively transported to antigen presenting cells. Particle size, surface charge, and vesicle disaggregation should be considered to ensure the efficacy and safety of the formulations. Griffiths *et al.* demonstrated that intratracheal instillation of liposomally encapsulated ricin toxoid vaccine produced high titer of ricin-specific antibodies than formaldehyde-treated ricin vaccine and alhydrogel vaccine solution [119]. Rebecca *et al.* studied the affect of polylactides-co-glycolide (PLG) microparticles on the antigen uptake and development of immune response after DNA vaccination in mice. Results from this study concluded that PLG formulation significantly reduced lung eosinophilia and boosted $CD8^{+}T$ cells while weakening the Th1 responses [120]. Vaccines targeting the antigens specific to lung cancer improved tumor shrinkage response and survival of NSCLC patients. Liposomal vaccine containing MUC1 lipopeptide (L-BPL25) is in clinical trials for the treatment of NSCLC [121].

3.3.8 Nucleic Acids

Gene therapy has been attempted to treat airway diseases, such as asthma, chronic obstructive pulmonary disease, cystic fibrosis, lung cancer, and pulmonary hypertension. Table 3.5 illustrates the various pulmonary gene formulations and their references. Also, gene delivery has been applied to modulate airway innate immune responses that are involved in cytokine imbalance and airway damage. Aerosolized gene therapy delivers a high dose to its target site with fewer side effects than intravenous administration. However, this therapy has met with limited success due to problems associated with targeting, low gene transfer efficiency, development of immune, and inflammatory responses [122]. Target defective genes have been either replaced or inhibited to modify the gene expression. Aerosolized gene therapy has been extensively investigated to study the transfer of cystic fibrosis transmembrane conductance regulator gene. Genes (CFTR) can be incorporated into viral and nonviral vectors. Nonviral vectors such as cationic liposomes, glycoconjugates, and DNA complexes with cationic lipids or polymers have been utilized for gene transfer. Cationic liposomes and glycoconjugates have been investigated for the delivery of CFTR gene, but these strategies were not particularly successful due to high toxicity and low transfection efficiency. Nuclear delivery has been facilitated with the attachment of nuclear localization signals and peptide sequences that facilitate cell internalization and endosomal escape. Gene replacement therapy has been investigated in patients with cystic fibrosis. CFTR gene has been inserted into a replication defective adenovirus vector. However, the transfection efficiency is low and associated with host immune responses against adenovirus vector [123]. Viral vectors such as lentiviruses, adeno-associated viruses, and adenoviruses have been targeted to deliver therapeutic genes. However, success has been limited due to immunogenicity and ontogenetic potential [124]. Adenoviruses were initially considered ideal viral vectors for gene transfer to the airway epithelium. Later on, it was concluded that adenoviruses delivered to human bronchial epithelium were ineffective and caused an acute inflammatory response. Several *in vitro* and *in vivo* models have been developed to investigate the efficiency of gene transfection and host responses to gene delivery vehicles [125]. The delivery of prostaglandin G/H synthetase gene and nitric oxide synthetase via liposomes has been investigated for the treatment of pulmonary hypertension. Anti-inflammatory genes such as superoxide dismutase enclosed in adenoviral vector have been shown to ameliorate lung injury in the mouse model following intratracheal instillation. Host immune responses to vectors and physical barriers (surfactant proteins and mucus) are major obstacles for the success of gene therapy. Immune responses are muted by improving vector design as well as using helper-dependent adenoviral vector (HDAd) systems. HDAd vectors are third-generation adenoviral vectors, also termed as gutless vectors because of the deletion of virus encoding sequences.

Table 3.5 Selected pulmonary gene delivery formulations reported in literature

Gene	Formulation	Animal model	Year and reference
Human cystic fibrosis transmembrane regulator (CFTR) gene	Recombinant adenovirus-associated vector	Humans	Moss *et al.* 2007 [126]
Programmed cell death receptor 4 (PCDR4)	Glucosylated polyethylenimine (GPEI)	Humans	Hwang *et al.* 2007 [127]
CFTR gene	PEO-PPO-PEO polymeric micelles	Nude mice	Chao *et al.* 2007 [128]
Plasmid DNA	Cationic polymer polyethylenimine	BALB/c and NMRI, and mixed 129/Sv x C57BL/6 mice	Dames *et al.* 2007 [129]
Interleukin-10	Adeno-associated virus vector	Murine	Fu *et al.* 2007 [130]
PCDR4 gene	Urocanic acid–modified chitosan (UAC)	K-ras null lung cancer mice model	Jin *et al.* 2006 [131]
Plasmid DNA	Polyethylenimine	Mice	Rudolph *et al.* 2005 [132]
Manganese superoxide dismutase plasmid (MnSOD)	Liposome	Mice	Carpenter *et al.* 2005 [133]
Alpha-1-antitrypsin (A1A4)	Recombinant adeno-associated virus	C57BL/6J mice	Zhang *et al.* 2003 [134]
P53 DNA	polyethylenimine	C57BL/6 mice	Koshkina *et al.* 2003 [135]

These vectors have demonstrated an improvement in transfection efficiency and reduced toxicity. Mucus barrier can be overcome by treatment with mucolytic agents such as *N*-acetyl-L-cysteine, leading to increased transfection efficiency with cationic liposomes. Several approaches such as intravenous, intranasal, and intratracheal have been studied for gene delivery to lungs. Rudolph *et al.* have shown that intratracheal instillation of aerosolized DNA complexed to PEI led to increase in gene expression after 24 hours [132]. Dames *et al.* have shown consistent gene expression levels in murine lungs for nine days after aerosolized delivery of PEI 25 kD/DNA [129]. While the intratracheal instillation is not particularly feasible in human, inhalation aerosol delivery can enhance transgene expression with lower toxicity. Cationic polypeptides such as polylysine, polyethyleneimine, protamines, and histones have been utilized for the treatment of lung cancer. Polyethyleneimine has been applied extensively for aerosol delivery due to its stability during nebulization. Aerosol polyethyleneimine based p53 and interleukin-12 therapy can significantly reduce the size of osteosarcoma-based lung metastases. Chitosan has been used as a nontoxic alternative to other cationic polymers. Hoggard *et al.* studied the effectiveness of

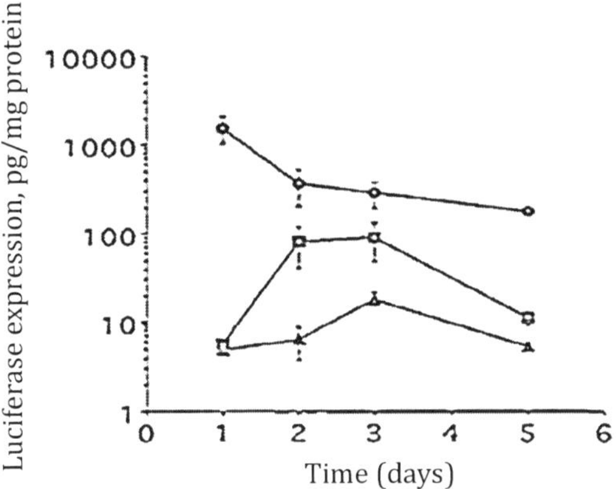

Figure 3.8 Luciferase gene expression in mouse lungs 1, 2, 3, and 5 days after intratracheal administration of 50 mg pCMV-Luc. UPC (solid rectangle) and PEI 25 KDa (solid circle) polyplexes were formulated at charge ratio 3:1 and 5:1 (-/-), respectively. Unformulated, naked DNA (solid triangle) was used as control. Results are expressed as mean values ≠ s.d. from one representative experiment (n = 4) of two performed.

chitosan formulation as a nonviral gene delivery system. As depicted in Fig. 3.8, this study demonstrated that chitosan has transfection efficiency that is comparable to PEI in inducing luciferase gene expression in mouse lungs after intratracheal instillation.

3.3.9 Oligonucleotides

Antisense and siRNA oligonucleotides hold great promise in silencing the gene expression with higher specificity and efficacy. siRNA is capable of 70–100% silencing of a specific gene with higher binding affinity, greater stability, and lower toxicity. Complex and branched anatomy of lungs, mucus layer, and airway cell membrane permeability are main barriers to the successful delivery of oligonucleotides to lungs. For efficient delivery, oligonucleotides should be stable, be targeted to site of action, readily cross the plasma membrane, and perform their activity with lower toxicity. Initially, cationic liposomes were selected for pulmonary delivery as these compounds readily form complexes with oligonucleotides through electrostatic interactions. Pretreatment of mice treated with cationic lipid vectors containing antisense oligonucleotides targeted to Inter-Cellular Adhesion Molecule 1(ICAM-1) gene significantly reduced ICAM-1 expression in lungs. However, cationic liposomes cause several disadvantages such as low encapsulation efficiency and poor ionic interactions. Recently, oligonucleotides entrapped in a cationic lipid/ODN complex formed by reverse evaporation method were developed by Stuart *et al.* [136]. DODAP/ODN complexes increased the entrapment efficiency of oligonucleotides by 60–80%. These complexes are further modified with antibodies to enhance the specificity. Endothelium-specific antibody to platelet/endothelial cell adhesion molecule-1 (PECAM-1) was recently complexed to oligonucleotide-encapsulated lipid vesicles in order to increase the entrapment efficiency and specificity. Several *in vitro* and *in vivo* models have been examined to study the transfection efficiency and cytotoxicity [137]. Cationic polymers such as polyethylenimine combined with siRNA have been shown to lower the influenza virus titers in the mouse lung [138]. Inhalable siRNA has been delivered to treat cystic fibrosis, infectious disease, and cancer. Ren *et al.* investigated the siRNA-mediated silencing of c-erbB-2 oncogene in Calu-3 cells [139]. Zhang *et al.* found that adenovirus-mediated siRNA can significantly reduce the mutant k-ras protein in lung cancer cell lines [140]. For treatment of cystic fibrosis, siRNA has been known to reduce inflammatory interleukin-8 levels. SiRNA Inc developed anti IL-4 siRNA for lung delivery with 80% decrease in airway responsiveness *in vivo*. Inhaled siRNA for the treatment of respiratory syncytial virus developed by Alnylam in 2006 is currently in clinical trials.

3.4 Conclusions

Patient friendly, noninvasive delivery systems have shown great promise for the treatment of chronic diseases such as diabetes, which require repeated injections. The unique chemical and physical properties of macromolecules and the complex physiology of the respiratory tract pose several challenges for nasal and pulmonary delivery. With the recent advances in delivery devices and biotechnology, a great potential exists in the development of novel nasal and pulmonary delivery systems intended for systemic administration of macromolecules. Particulate nanocarriers are able to improvise the physicochemical properties and sustain the release of the formulation, thereby prolonging the desired pharmacological or immune responses. Furthermore, targeted delivery of these bioactive macromolecules by attaching surface ligands, antibodies, and cell-penetrating peptides has been pursued to enhance the efficiency of the formulation. Research in the field of vaccines, gene therapy, and antisense oligonucleotides has generated significant breakthroughs in the treatment of life-threatening diseases such as cancer.

References

1. Goldberg, M., and I. Gomez-Orellana. (2003). Challenges for the oral delivery of macromolecules, *Nat Rev Drug Discov*. 2, pp. 289–295.
2. Gomez-Orellana, I. (2005). Strategies to improve oral drug bioavailability, *Expert Opin Drug Deliv*. 2, pp. 419–433.
3. Costantino, H.R. *et al*. (2007). Intranasal delivery: physicochemical and therapeutic aspects, *Int J Pharm*. 337, pp. 1–24.
4. Turker, S., E. Onur, and Y. Ozer. (2004). Nasal route and drug delivery systems, *Pharm World Sci*. 26, pp. 137–142.

5. Varricchio, A. *et al.* (2010). The nose and paranasal sinuses, *Int J Immunopathol Pharmacol.* 23, pp. 1–3.
6. Chien, Y.W., and S.-F. Chang. (1989). Drugs and the pharmaceutical sciences series. *Informa Healthcare*, vol. 39.
7. Lisbeth, I. (1999). *Bioadhesive Formulations for Nasal Peptide Delivery: Fundamentals, Novel Approaches and Development*, vol. 98. (ed. C.-M.L. Edith Mathiowitz and D.E. Chickering), Marcel Dekker, New York, 507–539.
8. Dodane, V., M.A. Khan, and J.R. Merwin. (1999). Effect of chitosan on epithelial permeability and structure, *Int J Pharm.* 182, pp. 21–32.
9. Ozsoy, Y., S. Gungor, and E. Cevher. (2009). Nasal delivery of high molecular weight drugs, *Molecules.* 14, pp. 3754–3779.
10. Lang, S. *et al.* (1996). Transport and metabolic pathway of thymocartin (TP4) in excised bovine nasal mucosa, *J Pharm Pharmacol.* 48, pp. 1190–1196.
11. Arora, P., S. Sharma, and S. Garg. (2002). Permeability issues in nasal drug delivery, *Drug Discov Today.* 7, pp. 967–975.
12. Marom, Z., J. Shelhamer, and M. Kaliner. (1984). Nasal mucus secretion, *Ear Nose Throat J.* 63, pp. 36–37, 41–42, 44.
13. Jones, N. (2001). The nose and paranasal sinuses physiology and anatomy, *Adv Drug Deliv Rev.* 51, pp. 5–19.
14. Agarwal, V., and B. Mishra. (1999). Recent trends in drug delivery systems: intranasal drug delivery, *Indian J Exp Biol.* 37, pp. 6–16.
15. Sarkar, M.A. (1992). Drug metabolism in the nasal mucosa, *Pharm Res.* 9, pp. 1–9.
16. Huang, Y., and M.D. Donovan. (1998). Large molecule and particulate uptake in the nasal cavity: the effect of size on nasal absorption, *Adv Drug Deliv Rev.* 29, pp. 147–155.
17. Jadhav, K.R., I.M.S. Manoj, N. Gambhire, V.J. Kadam, and S.S. Pisal. (2007). Nasal drug delivery system-factors affecting and applications, *Current Drug Ther.* 2, pp. 27–38.
18. Lipworth, B.J., and C.M. Jackson. (2000). Safety of inhaled and intranasal corticosteroids: lessons for the new millennium, *Drug Saf.* 23, pp. 11–33.
19. Huang, C.H. *et al.* (1985). Mechanism of nasal absorption of drugs I: Physicochemical parameters influencing the rate of in situ nasal absorption of drugs in rats, *J Pharm Sci.* 74, pp. 608–11.
20. Ohwaki, T. *et al.* (1985). Effects of dose, pH, and osmolarity on nasal absorption of secretin in rats, *J Pharm Sci.* 74, pp. 550–552.

21. Pujara, C.P., Z.S. Michelle, R. Duncan, and A.K. Mitra. (1995). Effects of formulation variables on nasal epithelial cell integrity: biochemical evaluations, *Int J Pharm.* 114, pp. 197–203.
22. Hardy, J.G., S.W. Lee, and C.G. Wilson. (1985). Intranasal drug delivery by spray and drops, *J Pharm Pharmacol.* 37, pp. 294–297.
23. Merkus, F.W. *et al.* (1991). Absorption enhancing effect of cyclodextrins on intranasally administered insulin in rats, *Pharm Res.* 8, pp. 588–592.
24. Lee, W.A. *et al.* (1994). The bioavailability of intranasal salmon calcitonin in healthy volunteers with and without a permeation enhancer, *Pharm Res.* 11, pp. 747–750.
25. Laursen, T. *et al.* (1994). Nasal absorption of growth hormone in normal subjects: studies with four different formulations, *Ann Pharmacother.* 28, pp. 845–848.
26. Agu, R.U. *et al.* (2002). Nasal absorption enhancement strategies for therapeutic peptides: an in vitro study using cultured human nasal epithelium, *Int J Pharm.* 237, pp. 179–191.
27. Karla, P.K., D.K. Ripal Gaudana, and A.K. Mitra. (2008). Drugs and the pharmaceutical sciences series, 2nd edn. *Informa Healthcare*, J.H. Michael, J. Rathbone, M.S. Roberts, and M.E. Lane, vol. 2. p. 728.
28. Francis, G. *et al.* (2009). Intranasal insulin ameliorates experimental diabetic neuropathy, *Diabetes.* 58, pp. 934–945.
29. Khafagy el, S. *et al.* (2009). Effect of cell-penetrating peptides on the nasal absorption of insulin, *J Controlled Release.* 133, pp. 103–108.
30. Jain, A.K. *et al.* (2007). Muco-adhesive multivesicular liposomes as an effective carrier for transmucosal insulin delivery, *J Drug Target.* 15, pp. 417–427.
31. McInnes, F.J. *et al.* (2007). Nasal residence of insulin containing lyophilised nasal insert formulations, using gamma scintigraphy, *Eur J Pharm Sci.* 31, pp. 25–31.
32. Wang, J., Y. Tabata, and K. Morimoto. (2006). Aminated gelatin microspheres as a nasal delivery system for peptide drugs: evaluation of in vitro release and in vivo insulin absorption in rats, *J Controlled Release.* 113, pp. 31–37.
33. Krauland, A.H. *et al.* (2006). In vivo evaluation of a nasal insulin delivery system based on thiolated chitosan, *J Pharm Sci.* 95, pp. 2463–2472.
34. Simon, M. *et al.* (2005). Insulin containing nanocomplexes formed by self-assembly from biodegradable amine-modified poly(vinyl alcohol)-graft-poly(L-lactide): bioavailability and nasal tolerability in rats, *Pharm Res.* 22, pp. 1879–1886.

35. Varshosaz, J., H. Sadrai, and R. Alinagari. (2004). Nasal delivery of insulin using chitosan microspheres, *J Microencapsul.* 21, pp. 761–774.

36. Lansley, A.B., and G.P. Martin. (2001). Nasal drug delivery, in *Drug Delivery and Targeting for Pharmacists and Pharmaceutical Scientists* (ed. A.W. Lloyd. A.M. Hillery, and J. Swarbrick), Taylor & Francis, New York, pp. 237–268.

37. Ishikawa, F. *et al.* (2001). Improved nasal bioavailability of elcatonin by insoluble powder formulation, *Int J Pharm.* 224, pp. 105–114.

38. Nagai T., N.Y. Nambu *et al.* (1984). Powder dosage of insulin for nasal administration, *J Controlled Release.* 1, pp. 15–22.

39. Chang, S-.F., and Y.W. Chien. (1991). Nasal drug delivery, in *Treatise on Controlled Drug Delivery: Fundamentals, Optimization, Applications* (ed. A. Kydonieus), Marcel Dekker, New York, pp. 423–464.

40. Teshima, D. *et al.* (2002). Nasal glucagon delivery using microcrystalline cellulose in healthy volunteers, *Int J Pharm.* 233, pp. 61–66.

41. Kararli, T.T. *et al.* (1992). Enhancement of nasal delivery of a renin inhibitor in the rat using emulsion formulations, *Pharm Res.* 9, pp. 1024–1028.

42. Varshosaz, J., H. Sadrai, and A. Heidari. (2006). Nasal delivery of insulin using bioadhesive chitosan gels, *Drug Deliv.* 13, pp. 31–38.

43. Law, S.L., K.J. Huang, and H.Y. Chou. (2001). Preparation of desmopressincontaining liposomes for intranasal delivery, *J Controlled Release.* 70, pp. 375–382.

44. Chen, M. *et al.* (2009). Improved absorption of salmon calcitonin by ultraflexible liposomes through intranasal delivery, *Peptides.* 30, pp. 1288–1295.

45. Ishii, M., and N. Kojima. (2010). Mucosal adjuvant activity of oligomannose-coated liposomes for nasal immunization, *Glycoconj J.* 27, pp. 115–123.

46. Iwanaga, K. *et al.* (2000). Usefulness of liposomes as an intranasal dosage formulation for topical drug application, *Biol Pharm Bull.* 23, pp. 323–326.

47. Vila, A. *et al.* (2005). *Int J Pharm.* 292, pp. 43–52.

48. Jain, A.K. *et al.* (2008). Effective insulin delivery using starch nanoparticles as a potential trans-nasal mucoadhesive carrier, *Eur J Pharm Biopharm.* 69, pp. 426–435.

49. Fernandez-Urrusuno, R. *et al.* (1999). Enhancement of nasal absorption of insulin using chitosan nanoparticles, *Pharm Res.* 16, pp. 1576–1581.

50. Wang, X. *et al.* (2009). Chitosan-NAC nanoparticles as a vehicle for nasal absorption enhancement of insulin, *J Biomed Mater Res B Appl Biomater*. 88, pp. 150–161.
51. Bhumkar, D.R. *et al.* (2007). Chitosan reduced gold nanoparticles as novel carriers for transmucosal delivery of insulin, *Pharm Res*. 24, pp. 1415–1426.
52. Zhang, X. *et al.* (2008). Nasal absorption enhancement of insulin using PEG-grafted chitosan nanoparticles, *Eur J Pharm Biopharm*. 68, pp. 526–534.
53. Teijeiro-Osorio, D., C. Remunan-Lopez, and M.J. Alonso. (2009). New generation of hybrid poly/oligosaccharide nanoparticles as carriers for the nasal delivery of macromolecules, *Biomacromolecules*. 10, pp. 243–249.
54. Illum L., H. Jorgensen, H. Bisgaard, O. Krogsgaard, and N. Rossing. (1987). Bioadhesive microspheres as a potential nasal drug delivery system, *Int J Pharm*. 39, pp. 189–199.
55. Illum, L. *et al.* (2001). Bioadhesive starch microspheres and absorption enhancing agents act synergistically to enhance the nasal absorption of polypeptides, *Int J Pharm*. 222, pp. 109–119.
56. Björk E., and P. Edman. (1990). Characterization of degradable starch microspheres as a nasal delivery system for drugs, *Int J Pharm*. 62, pp. 187–192.
57. Holmberg, K. *et al.* (1994). Influence of degradable starch microspheres on the human nasal mucosa, *Rhinology*. 32, pp. 74–77.
58. Romeo, V.D. *et al.* (1998). Effects of physicochemical properties and other factors on systemic nasal drug delivery, *Adv Drug Deliv Rev*. 29, pp. 89–116.
59. Illum, L., N.F. Farraj, S.S. Davis, B.R. Johansen, and D.T. O'Hagan. (1990). Investigation of the nasal absorption of biosynthetic human growth hormone in sheep — use of a bioadhesive microsphere delivery system, *Int J Pharm*. 63, pp. 207–211.
60. Critchley, H. *et al.* (1994). Nasal absorption of desmopressin in rats and sheep. Effect of a bioadhesive microsphere delivery system, *J Pharm Pharmacol*. 46, pp. 651–656.
61. Jung, T. *et al.* (2001). Tetanus toxoid loaded nanoparticles from sulfobutylated poly(vinyl alcohol)-graft-poly(lactide-co-glycolide): evaluation of antibody response after oral and nasal application in mice, *Pharm Res*. 18, pp. 352–360.
62. Illum, L. *et al.* (2001). Chitosan as a novel nasal delivery system for vaccines, *Adv Drug Deliv Rev*. 51, pp. 81–96.

63. Singh, J. *et al.* (2006). Diphtheria toxoid loaded poly-(epsilon-caprolactone) nanoparticles as mucosal vaccine delivery systems, *Methods.* 38, pp. 96–105.

64. Sayin, B. *et al.* (2009). TMC-MCC (*N*-trimethyl chitosan-mono-*N*-carboxymethyl chitosan) nanocomplexes for mucosal delivery of vaccines, *Eur J Pharm Sci.* 38, pp. 362–369.

65. Anwer, K., B.G. Rhee, and S.K. Mendiratta. (2003). Recent progress in polymeric gene delivery systems, *Crit Rev Ther Drug Carrier Syst.* 20, pp. 249–293.

66. Jia, S.F. *et al.* (2002). Eradication of osteosarcoma lung metastases following intranasal interleukin-12 gene therapy using a nonviral polyethylenimine vector, *Cancer Gene Ther.* 9, pp. 260–266.

67. Hyde, S.C. *et al.* (2000). Repeat administration of DNA/liposomes to the nasal epithelium of patients with cystic fibrosis, *Gene Ther.* 7, pp. 1156–1165.

68. Kumar, M. *et al.* (2002). Intranasal gene transfer by chitosan-DNA nanospheres protects BALB/c mice against acute respiratory syncytial virus infection, *Hum Gene Ther.* 13, pp. 1415–1425.

69. Khatri, K. *et al.* (2008). Surface modified liposomes for nasal delivery of DNA vaccine, *Vaccine.* 26, pp. 2225–2233.

70. Lee, D. *et al.* (2007). Thiolated chitosan/DNA nanocomplexes exhibit enhanced and sustained gene delivery, *Pharm Res.* 24, pp. 157–167.

71. Zhou, S. *et al.* (2010). Intranasal administration of CpG DNA lipoplex prevents pulmonary metastasis in mice, *Cancer Lett.* 287, pp. 75–81.

72. Csaba, N. *et al.* (2009). Ionically crosslinked chitosan nanoparticles as gene delivery systems: effect of PEGylation degree on in vitro and in vivo gene transfer, *J Biomed Nanotechnol.* 5, pp. 162–171.

73. Sundaram, S. *et al.* (2009). Surface-functionalized nanoparticles for targeted gene delivery across nasal respiratory epithelium, *FASEB J.* 23, pp. 3752–3765.

74. Wang, S.Y. *et al.* (2008). Intranasal delivery of T-bet modulates the profile of helper T cell immune responses in experimental asthma, *J Investig Allergol Clin Immunol.* 18, pp. 357–365.

75. Buckley, S.M. *et al.* (2008). Luciferin detection after intranasal vector delivery is improved by intranasal rather than intraperitoneal luciferin administration, *Hum Gene Ther.* 19, pp. 1050–1056.

76. Gomes, D.C. *et al.* (2007). Intranasal delivery of naked DNA encoding the LACK antigen leads to protective immunity against visceral leishmaniasis in mice, *Vaccine.* 25, pp. 2168–2172.

77. Kim, T.W. *et al.* (2005). Airway gene transfer using cationic emulsion as a mucosal gene carrier, *J Gene Med.* 7, pp. 749–758.

78. Tanaka, S.I. *et al.* (2004). Daily nasal inoculation with the insulin gene ameliorates diabetes in mice, *Diabetes Res Clin Pract.* 63, pp. 1–9.

79. Suarez, S., and A.J. Hickey. (2000). Drug properties affecting aerosol behavior, *Respir Care.* 45, pp. 652–666.

80. Siekmeier, R., and G. Scheuch. (2008). Systemic treatment by inhalation of macromolecules--principles, problems, and examples, *J Physiol Pharmacol.* 59 Suppl 6, pp. 53–79.

81. Yamamoto, A., S. Umemori, and S. Muranishi. (1994). Absorption enhancement of intrapulmonary administered insulin by various absorption enhancers and protease inhibitors in rats, *J Pharm Pharmacol.* 46, pp. 14–18.

82. Smola, M., T. Vandamme, and A. Sokolowski. (2008). Nanocarriers as pulmonary drug delivery systems to treat and to diagnose respiratory and non respiratory diseases, *Int J Nanomedicine.* 3, pp. 1–19.

83. Guthi, J.S. *et al.* MRI-visible micellar nanomedicine for targeted drug delivery to lung cancer cells, *Mol Pharm.* 7, pp. 32–40.

84. Torchilin, V.P. *et al.* (2003). Immunomicelles: targeted pharmaceutical carriers for poorly soluble drugs, *Proc Natl Acad Sci USA.* 100, pp. 6039–6044.

85. Lu, D., and A.J. Hickey. (2005). Liposomal dry powders as aerosols for pulmonary delivery of proteins, *AAPS PharmSciTech.* 6, pp. E641–E648.

86. Bi, R. *et al.* (2008). Spray-freeze-dried dry powder inhalation of insulin-loaded liposomes for enhanced pulmonary delivery, *J Drug Target.* 16, pp. 639–648.

87. Bruck, A. *et al.* (2001). Lectin-functionalized liposomes for pulmonary drug delivery: interaction with human alveolar epithelial cells, *J Drug Target.* 9, pp. 241–251.

88. Khalil, I.A. *et al.* (2007). Octaarginine-modified multifunctional envelope-type nanoparticles for gene delivery, *Gene Ther.* 14, pp. 682–689.

89. Zou, Y. *et al.* (2000). Development of cationic liposome formulations for intratracheal gene therapy of early lung cancer, *Cancer Gene Ther.* 7, pp. 683–696.

90. Gautam, A., J.C. Waldrep, and C.L. Densmore. (2003). Aerosol gene therapy, *Mol Biotechnol.* 23, pp. 51–60.

91. Garbuzenko, O.B. *et al.* (2009). Intratracheal versus intravenous liposomal delivery of siRNA, antisense oligonucleotides and anticancer drug, *Pharm Res.* 26, pp. 382–394.

92. Colonna, C. *et al.* (2008). Non-viral dried powders for respiratory gene delivery prepared by cationic and chitosan loaded liposomes, *Int J Pharm.* 364, pp. 108–118.

93. Yang, Y. *et al.* (2009). Development of highly porous large PLGA microparticles for pulmonary drug delivery, *Biomaterials.* 30, pp. 1947–1953.

94. Edwards, D.A. *et al.* (1997). Large porous particles for pulmonary drug delivery, *Science.* 276, pp. 1868–1871.

95. Edwards, D.A., A. Ben-Jebria, and R. Langer. (1998). Recent advances in pulmonary drug delivery using large, porous inhaled particles, *J Appl Physiol.* 85, pp. 379–385.

96. Mansour, H.M., Y.S. Rhee, and X. Wu. (2009). Nanomedicine in pulmonary delivery, *Int J Nanomedicine.* 4, pp. 299–319.

97. Grenha, A., B. Seijo, and C. Remunan-Lopez. (2005). Microencapsulated chitosan nanoparticles for lung protein delivery, *Eur J Pharm Sci.* 25, pp. 427–437.

98. Elfinger, M. *et al.* (2009). Self-assembly of ternary insulin-polyethylenimine (PEI)-DNA nanoparticles for enhanced gene delivery and expression in alveolar epithelial cells, *Biomacromolecules.* 10, pp. 2912–2920.

99. Sommerville, M.L., and A.J. Hickey. (2003). Aerosol generation by metered-dose inhalers containing dimethyl ether/propane inverse microemulsions, *AAPS PharmSciTech.* 4, p. E58.

100. Hamishehkar, H. *et al.* (2010). Pharmacokinetics and pharmacodynamics of controlled release insulin loaded PLGA microcapsules using dry powder inhaler in diabetic rats, *Biopharm Drug Dispos.* 31, pp. 189–201.

101. Bi, R. *et al.* (2009). Solid lipid nanoparticles as insulin inhalation carriers for enhanced pulmonary delivery, *J Biomed Nanotechnol.* 5, pp. 84–92.

102. Lim, S.H. *et al.* (2009). Human insulin microcrystals with lactose carriers for pulmonary delivery, *Biosci Biotechnol Biochem.* 73, pp. 2576–2582.

103. Zhang, Y. *et al.* (2009). The preparation and application of pulmonary surfactant nanoparticles as absorption enhancers in insulin dry powder delivery, *Drug Dev Ind Pharm.* 35, pp. 1059–1065.

104. Chono, S. *et al.* (2009). Aerosolized liposomes with dipalmitoyl phosphatidylcholine enhance pulmonary insulin delivery, *J Controlled Release.* 137, pp. 104–109.

105. Patel, L.N. *et al.* (2009). Conjugation with cationic cell-penetrating peptide increases pulmonary absorption of insulin, *Mol Pharm.* 6, pp. 492–503.

106. Ungaro, F. *et al.* (2009). Insulin-loaded PLGA/cyclodextrin large porous particles with improved aerosolization properties: in vivo deposition and hypoglycaemic activity after delivery to rat lungs, *J Controlled Release.* 135, pp. 25–34.

107. Liu, J. *et al.* (2008). Solid lipid nanoparticles for pulmonary delivery of insulin, *Int J Pharm.* 356, pp. 333–344.

108. Park, S.H. *et al.* (2007). Characterization of human insulin microcrystals and their absorption enhancement by protease inhibitors in rat lungs, *Int J Pharm.* 339, pp. 205–212.

109. Huang, Y.Y., and C.H. Wang. (2006). Pulmonary delivery of insulin by liposomal carriers, *J Controlled Release.* 113, pp. 9–14.

110. Rave, K. *et al.* (2004). Inhaled micronized crystalline human insulin using a dry powder inhaler: dose-response and time-action profiles, *Diabet Med.* 21, pp. 763–768.

111. Heijerman, H. *et al.* (2009). Inhaled medication and inhalation devices for lung disease in patients with cystic fibrosis: a European consensus, *J Cyst Fibros.* 8, pp. 295–315.

112. Frijlink, H.W., and A.H. De Boer. (2004). Dry powder inhalers for pulmonary drug delivery, *Expert Opin Drug Deliv.* 1, pp. 67–86.

113. Newman, S.P. (2005). Principles of metered-dose inhaler design, *Respir Care.* 50, pp. 1177–1190.

114. Chougule, M.B. *et al.* (2007). Development of dry powder inhalers, *Recent Pat Drug Deliv Formul.* 1, pp. 11–21.

115. Lu, D., and A.J. Hickey. (2007). Pulmonary vaccine delivery, *Expert Rev Vaccines.* 6, pp. 213–226.

116. Cutts, F.T., C.J. Clements, and J.V. Bennett. (1997). Alternative routes of measles immunization: a review, *Biologicals.* 25, pp. 323–338.

117. Jemski, J.V., and J.S. Walker. (1976). Aerosol vaccination of mice with a live, temperature- sensitive recombinant influenza virus, *Infect Immun.* 13, pp. 818–824.

118. Schwarz, L.A. *et al.* (1996). Delivery of DNA-cationic liposome complexes by small-particle aerosol, *Hum Gene Ther.* 7, pp. 731–741.

119. Griffiths, G.D. *et al.* (1997). Liposomally-encapsulated ricin toxoid vaccine delivered intratracheally elicits a good immune response and protects against a lethal pulmonary dose of ricin toxin, *Vaccine.* 15, pp. 1933–1939.

120. Helson, R. *et al.* (2008). Polylactide-co-glycolide (PLG) microparticles modify the immune response to DNA vaccination, *Vaccine.* 26, pp. 753–761.

121. Nemunaitis, J. (2007). A review of vaccine clinical trials for non-small cell lung cancer, *Expert Opin Biol Ther.* 7, pp. 89–102.

122. Rubin, B.K. (1999). Emerging therapies for cystic fibrosis lung disease, *Chest.* 115, pp. 1120–1126.

123. Knowles, M.R. *et al.* (1995). A controlled study of adenoviral-vector-mediated gene transfer in the nasal epithelium of patients with cystic fibrosis, *N Engl J Med.* 333, pp. 823–831.

124. Di Gioia, S., and M. Conese. (2009). Polyethylenimine-mediated gene delivery to the lung and therapeutic applications, *Drug Des Devel Ther.* 2, pp. 163–188.

125. Kushwah, R., H. Cao, and J. Hu. (2007). Potential of helper-dependent adenoviral vectors in modulating airway innate immunity, *Cell Mol Immunol.* 4, pp. 81–89.

126. Moss, R.B. *et al.* (2007). Repeated aerosolized AAV-CFTR for treatment of cystic fibrosis: a randomized placebo-controlled phase 2B trial, *Hum Gene Ther.* 18, pp. 726–732.

127. Hwang, S.K. *et al.* (2007). Aerosol-delivered programmed cell death 4 enhanced apoptosis, controlled cell cycle and suppressed AP-1 activity in the lungs of AP-1 luciferase reporter mice, *Gene Ther.* 14, pp. 1353–1361.

128. Chao, Y.C. *et al.* (2007). Ethanol enhanced in vivo gene delivery with non-ionic polymeric micelles inhalation, *J Controlled Release.* 118, pp. 105–117.

129. Dames, P. *et al.* (2007). Aerosol gene delivery to the murine lung is mouse strain dependent, *J Mol Med.* 85, pp. 371–378.

130. Fu, C.L. *et al.* (2006). Effects of adenovirus-expressing IL-10 in alleviating airway inflammation in asthma, *J Gene Med.* 8, pp. 1393–1399.

131. Jin, H. *et al.* (2006). Aerosol delivery of urocanic acid-modified chitosan/programmed cell death 4 complex regulated apoptosis, cell cycle, and angiogenesis in lungs of K-ras null mice, *Mol Cancer Ther.* 5, pp. 1041–1049.

132. Rudolph, C. *et al.* (2005). Aerosolized nanogram quantities of plasmid DNA mediate highly efficient gene delivery to mouse airway epithelium, *Mol Ther.* 12, pp. 493–501.

133. Carpenter, M. *et al.* (2005). Inhalation delivery of manganese superoxide dismutase-plasmid/liposomes protects the murine lung from irradiation damage, *Gene Ther.* 12, pp. 685–693.

134. Zhang, D. *et al.* (2003). Alpha-1-antitrypsin expression in the lung is increased by airway delivery of gene-transfected macrophages, *Gene Ther.* 10, pp. 2148–2152.

135. Koshkina, N.V. *et al.* (2003). Biodistribution and pharmacokinetics of aerosol and intravenously administered DNA-polyethyleneimine complexes: optimization of pulmonary delivery and retention, *Mol Ther.* 8, pp. 249–254.

136. Stuart, D.D., S.C. Semple, and T.M. Allen. (2004). High efficiency entrapment of antisense oligonucleotides in liposomes, *Methods Enzymol.* 387, pp. 171–188.

137. Wilson, A. *et al.* (2005). Targeted delivery of oligodeoxynucleotides to mouse lung endothelial cells in vitro and in vivo, *Mol Ther.* 12, pp. 510–518.

138. Ge, Q. *et al.* (2004). Inhibition of influenza virus production in virus-infected mice by RNA interference, *Proc Natl Acad Sci USA.* 101, pp. 8676–8681.

139. Ren, S.H. *et al.* (2005). [Effect of RNAi-mediated gene silencing of C-erbB-2 on proliferation of lung adenocarcinoma cell line calu-3], article in Chinese, *Ai Zheng.* 24, pp. 1173–1178.

140. Zhang, X. *et al.* (2004). Small interfering RNA targeting heme oxygenase-1 enhances ischemia-reperfusion-induced lung apoptosis, *J Biol Chem.* 279, pp. 10677–10684.

Chapter 4

In vitro and *in vivo* Diagnosis of Pulmonary Disorders Using Nanotechnology

Indrajit Roy
Department of Chemistry,
University of Delhi,
Delhi-110007, India
indrajitroy11@gmail.com

4.1 Introduction

Nanotechnology is playing a rapidly increasing role in the diagnostics and therapeutics of human diseases [1–5]. The integration of nanoscience and nanotechnology into biomedical research is ushering in a true revolution that is broadly impacting biotechnology. New terms such as "nanobioscience," "nanobiotechnology," and "nanomedicine" are gaining wide acceptance. Nanochemistry deals with the confinement of chemical reactions to produce nanometer-scale products (generally in the 1 to 100 nm size range). New materials and ideas, including carbon nanotubes (CNTs), ultrasensitive biosensors, smart materials, molecular motors, shape memory materials, and biological templates for nanoscale devices, are some of the recent products of research in nanotechnology.

The lung is an attractive target for nanoparticulate drug delivery owing to the availability of a large surface area [6–8]. In addition, this organ is accessible via the inhalation route, which avoids the

Pulmonary Nanomedicine: Diagnostics, Imaging, and Therapeutics
Edited by Neeraj Vij

ISBN 978-981-4316-48-4 (Hardback), 978-981-4364-14-0 (eBook)
www.panstanford.com

complications associated with systemic delivery to this organ, such as capture by the reticuloendothelial system (RES) [9, 10]. As a result, a variety of nanoparticles are under active investigation from the perspective of pulmonary delivery, which includes polymeric nanocarriers, liposomes, gold nanoparticles (GNPs), quantum dots (QDs), iron oxide nanoparticles, and CNTs [11–16]. The topics covered include the application of nanoparticles containing active agents for the diagnosis, therapeutics, and prophylaxis of various lung diseases. However, in conjunction with these benefits, there is also a growing concern of pulmonary toxicity of nanoparticles as a result of their inhalation/administration, either accidental or purposeful [17–20]. These concerns mandate a careful evaluation of the risks and benefits of the biomedical application of various nanoparticles in the lung.

This chapter focuses on the diagnostic aspects of nanoparticles from the point of view of early and accurate identification of various lung diseases. This can be achieved either by *in vitro* analysis of extracted pulmonary cells or isolated body fluids for the quantification of various biomarkers indicative of lung malignancies or by noninvasive imaging of diagnostic nanoparticles targeted to specific areas/cells within the lung *in vivo*. We will begin with a discussion on the nanoparticles that are better suited for pulmonary diagnostic purposes. We will follow with a discussion on the various techniques/tools for nanotechnology-mediated *in vitro* and *in vivo* diagnosis. We will then discuss the various lung malignancies where it is critical to develop novel methods for rapid and accurate diagnosis and disease stratification, by either *in vitro* or *in vivo* analysis, stating their scope, various challenges that they currently encounter, and how nanotechnology can overcome these challenges, along with some specific examples highlighting each case. We will finish with a discussion on the toxicological aspects of the various nanoparticles with regards to pulmonary delivery or exposure.

4.2 Nanoparticles

A number of nanoparticles are being currently investigated in biomedical applications. However, in the realm of diagnostics, inorganic nanoparticles are better suited over polymeric or liposomal nanoformulations owing to their structural robustness, inert nature, ease of storage and transportation, and resistance to microbial attack [21–24]. In addition, some of these inorganic nanomaterials, such as QDs/rods, GNPs, and iron oxide nanoparticles, have "in built" diagnostic signatures, which do away with the necessity of tagging external imaging probes to them [25, 26]. In the subsequent section, we will discuss the key features of some of these inorganic nanoparticles, which have already played or are expected to play a critical role in the diagnosis of pulmonary and other diseases, via *in vitro* and *in vivo* analysis.

4.2.1 Quantum Dots

QDs are semiconductor nanoparticles that have unique size-tunable optical properties spanning the ultraviolet (UV) to the near infrared (NIR) region of the spectrum [27–29]. They have numerous advantages over conventional organic fluorophores that make them ideal new-generation optical probes for ultrasensitive and multiplexed detection of various disease biomarkers. These include enhanced brightness, high photostability, broad excitation and narrow bandwidth emission spectra, and ease of surface bioconjugation. These properties enable the fabrication of numerous multicolored QD-based bioconjugates that can be simultaneously excited using a single light source, and their emission signals can be detected with minimal spectral overlap. Therefore, QD-based nanoprobes are gaining increasing attention in numerous bioanalytical and diagnostic techniques, which include microarray technology, genomics/proteomics, western blotting, quantitative polymerase chain reaction (PCR), enzyme-linked immunosorbent assay (ELISA), optical biosensors, and flow cytometry [30, 31]. In addition, QDs are also being extensively investigated for *in vivo* diagnosis and drug delivery, as some of them are reported to be safely excreted out of the body via the renal route [28]. Interestingly, the first report of *in vivo* targeting of QD nanocrystals involved the systemic delivery of QDs conjugated with lung-targeting peptides to specific sites within the lung [15].

4.2.2 Rare Earth-Doped Nanophosphors

Nanophosphors are ceramic nanomaterials in which rare earth (RE) ions are embedded in a crystalline matrix [32–35]. They absorb and emit in the UV, visible, and NIR spectral ranges. By the use of different RE dopants, a large variety of spectrally distinctive emissions can be obtained. The most interesting feature of luminescence spectra of RE-doped nanophosphors is their sharp spectral lines, which resemble the spectra found in the case of corresponding free ions. As in the case of QDs, nanophosphors are optically stable and resistant to photobleaching. The RE ions also demonstrate photoluminescence lifetimes in submillisecond or millisecond range, thus facilitating time-gated detection for autofluorescence suppression [36]. Finally, depending on the choice of RE dopants, nanophosphors provide the capability of upconverted emission, in which the emitted energy is higher than the absorbed energy. An interesting example is NIR-to-NIR upconversion, where both the absorption and emission wavelengths are in the NIR range, thus facilitating deep-tissue and background-free imaging [34, 35]. Owing to these unique optical properties, such materials are generating increasing attention within the scientific community for the fabrication of novel diagnostic nanoprobes.

4.2.3 Dye-Doped Silica/ORMOSIL Nanoparticles

Fluorescent dye-doped nanoparticles represent an important class of nanomaterials for optical bioimaging [37–39]. Prominent among these materials are silica and organically modified silica (ORMOSIL), which have several favorable properties such as optical transparency, nonantigenicity, and rich surface chemistry for facile bioconjugation [42–44]. Here, a number of fluorescent dye molecules can be contained within each silica/ORMOSIL particle, which protects them from photobleaching and prevents their interaction with the biological environment. Fluorescent nanoparticles can permeate across cell membranes, which make them suitable for subcellular imaging applications. The absorption and emission of these nanoparticle-based probes are determined by the properties of encapsulated fluorophores. An increase of the signal output from individual nanoparticles can be accomplished by increasing the loading of the dye, which leads to an increase in the absorption cross-section per nanoparticle-based probe.

4.2.4 Gold Nanoparticles

Nanometer-sized metallic particles such as gold have emerged as a new class of materials for applications ranging from physics to biology [42–44]. Gold colloids are well-known for their localized surface plasmon resonance (LSPR) properties, which originate from the collective oscillation of their electrons in response to optical excitation. The LSPR frequency of a particular metal colloid has been shown to depend on particle shape, composition, and refractive index of the surrounding medium, among many other factors [45, 46]. Metallic nanoparticles can be functionalized with biomolecules (e.g., small-molecule drugs, aptamers, peptides, and antibodies) for specific targeting of tumor cells and early detection of cancer [47–49]. GNPs are also extensively investigated in the area of *in vitro* diagnosis by virtue of their unique and tunable plasmonic signatures, which change color upon aggregation and deaggregation. In addition, as GNPs are known to be safe for biomedical use, their use as targeted *in vivo* diagnostic agents is fast gaining popularity. Giordano *et al.* have reported the targeted delivery of bioconjugated GNPs in the lung vasculature using a phage display approach, which constitutes a promising diagnostic platform for imaging cancer, emphysema, and other lung diseases [49].

4.2.5 Iron Oxide Nanoparticles

Owing to their high magnetic susceptibility, biocompatibility, and ease of surface biofunctionalization, iron oxide nanoparticles serve as excellent contrast agents for magnetic resonance imaging (MRI) [50–52]. These nanoparticles are already approved for use in clinical MRI (ferridex) [53]. In addition to *in vivo* MRI, magnetic biosensors based on these nanoparticles have been designed to detect a wide range of targets, including DNA/mRNA, proteins, enzymes, drugs, pathogens, and tumor cells [54–56].

4.2.6 Carbon Nanotubes

CNTs have several unique physical and catalytic properties, which include high electrical conductivity, chemical stability, and mechanical strength [57–58]. These make CNTs ideal for use in electrochemical sensors, which has already found applications in the detection of neurotransmitters, other proteins and peptides, small molecules such as glucose, and DNA. Different types of electrochemical methods are used in these sensors, including direct electrochemical detection with amperometry or voltammetry, indirect detection of an oxidation product using enzyme sensors, and detection of conductivity changes using CNT-field effect transistors (FETs) [59].

4.3 *In vitro* Diagnosis, Techniques, and Challenges

Body fluid samples collected from humans, which include biopsied cells, blood, urine, sweat, sputum, and bronchio-alveolar lavage (BAL), are a rich source of secreted biomolecules that carry significant information about the health and well-being of the body [60–62]. These samples are also potent indicators of any malfunctioning of the body and can play a critical role in the diagnosis and staging of several, if not all, human diseases. *In vitro* diagnosis is based on the quantitative analysis of these body fluids using modern-day bioanalytical techniques, such as mass spectrometry, flow cytometry, and various ELISA assays [63, 64]. Such capability to quantify these biomarkers is critically important in accurately predicting these diseases in humans as well as stratifying them into appropriate etiologies/stages. Such diagnosis would help clinicians in not only administering appropriate therapeutic regimens specific to the nature and stage of the disease but also monitoring the outcome of the therapy by simple *in vitro* analysis of body fluids. The following section will serve as an introduction to various *in vitro* diagnostic techniques, discussing their setup, current limitations, and the role of nanotechnology in overcoming these limitations.

4.3.1 Flow Cytometry

Flow cytometry is a technique of counting and examining microscopic particles, such as cells or bioconjugated microspheres, by suspending them in a stream of fluid and passing them by an electronic detection apparatus [65]. One of the most promising multiplexed bioanalytical techniques is the suspension bead array (SBA) assay, which is a flexible analyzer based on the principles of flow cytometry that enables the multiplexed detection of proteins in a single sample [66, 67]. This technique involves decorating differentially "colored" microbeads (microparticles) with suitable "capture" monoclonal antibodies (mAb). The microbeads are manufactured in sets such that each set has a unique proportion of fluorescent dyes, enabling the cytometer to distinguish one set from the other. By conjugating each mAb to a specific bead set, multiple analytes can be distinguished and quantified. The sample containing a mixture of analytes (antigens) is incubated with the microbeads, and the analytes bound to the beads are detected with a cocktail of biotinylated mAbs followed by reaction with a streptavidin-conjugated fluorophore. The ability of such an assay to detect "soluble" antigens found in body secretions (e.g., blood, saliva, urine, and sputum) in a "high throughput" manner has significant implications in the rapid identification of specific biomarkers associated with various diseases/disorders.

Unfortunately, the current state-of-the-art technique is fraught with numerous problems. First, the background interference of biologics in the samples impedes on the sensitivity. Second, the use of organic fluorophores as fluorescent probes leads to reduced signal-to-noise ratio due to their poor photostability [68, 69]. Third, the broad emission spectra of the fluorophores will result in considerable overlap in the emission profile when involving multicolor analysis, thus severely limiting the number of analytes that can be simultaneously detected [71]. Fourth, the use of multiple excitation wavelengths as excitation sources leads to severe interassay variability, further reducing sensitivity [71]. Therefore, a rapid, multiplexed, and ultrasensitive detection technique of various proteins in body fluids is urgently needed in order to facilitate the quantification of numerous biomarkers of various diseases/disorders at biologically and clinically relevant concentrations. In order to achieve these goals, the development of a new generation of detection probes is essential.

The unique optical properties of QDs can be exploited to fabricate a new generation of probes for flow cytometry, which can overcome the limitations of organic fluorophore-based probes [26–31]. Firstly, the high spectral and bench stability of the QD-based probes allow their continuous and recyclable use in biomarker detection. Next, their broad excitation spectra make possible the simultaneous excitation of multicolored QDs using a single laser excitation source, which not only leads to minimum interassay variability issues but also makes possible the fabrication of more compact, robust, and user friendly flow cytometers. In addition, their narrow emission spectra would facilitate multiplexed detection with minimum spectral overlap. Finally, the rich surface chemistry of QDs makes them useful for controlled bioconjugation with numerous antibodies, proteins, and nucleic acids, allowing the generalization of their use in the diagnosis of numerous diseases/disorders. Therefore, the development of QDs conjugated with suitable biorecognition molecules as probes in a high throughput bioanalytical device would allow rapid and "multiplexed" diagnosis of trace amounts of specific biomarkers indicative of a number of human diseases/disorders that are poorly diagnosed using currently existing technologies.

4.3.2 Multiplexed Microarray ELISA

Currently, many proteins are determined on the basis of their reactivity with specific antibodies, either in a radioimmunoassay format or using an enzyme linked immunosorbent assay (ELISA) [72]. Of these two methods, ELISAs are the most commonly used, and there are several commercial sources of kits or antibodies available that are suitable for these assays. Knight *et al.* have developed a rapid and sensitive method for measuring 16 human proteins in a microarray-based ELISA technique called microarray immunoassay, or MI [73]. The MI was constructed and processed using readily available equipment and commercially available reagents. Essentially, any laboratory with access to a microarray spotter and fluorescent glass slide reader may measure proteins using this methodology. The results are rapid and reproducible, and the sensitivity is equal to traditional ELISAs.

However, similar to flow cytometry as discussed above, the sensitivity, reproducibility, and multiplexibility of these assays are compromised as they use poorly photostable and broad-emitting organic fluorophores as optical probes. Therefore, they too need novel nanoparticulate optical probes with higher photostability and narrow emission, such as QDs, RE-doped nanophosphors, and dye-doped silica/ORMOSIL nanoparticles, in order to overcome these limitations [26–35].

4.3.3 Molecular Beacon Technology

Molecular beacon technology is an extremely popular method for the quantification of short RNA molecules, whether in solution or inside cells. A molecular beacon is an oligonucleotide labeled with a fluorophore and a quencher, which upon hybridization to a complementary target undergoes a conformational change, resulting in a fluorescent signal [74, 75]. These beacons are arranged as a hairpin with the target sequence in the loop and a noncomplementary sequence in the stem, with a fluorophore and its quencher attached to either ends of the oligonucleotide. In the native hairpin conformation, the quencher blocks the emission of the fluorophore as they are in close proximity. However, when the complementary target binds at the loop, the hairpin stem–loop structure is opened. This leads to the separation of the fluorophore and the quencher, resulting in a large increase in fluorescence signal.

The two major limitations of this technology in the current state are (a) limited sensitivity owing to the poor photostability of the organic fluorophores as probes and (b) poor intracellular delivery when considering imaging RNA within cells [76]. Nanoparticle-based probes can overcome both these limitations, for example by using QD-based probes for enhanced photostability and biotargeting molecules attached to nanoparticles for higher cellular uptake. GNPs have also been used as quenchers as they are known to effectively quench the fluorescence of QDs and organic fluorophores [77].

4.3.4 Plasmonic Biosensing

The unique property of LSPR associated with metallic nanoparticles paves the way toward the assembly of plasmonic biosensors [78, 79]. In particular, bioconjugates of GNPs have been extensively employed as plasmonic nanoprobes. This platform has found increased applications in the detection of a variety of biomolecular targets, including nucleic acids, proteins, saccharides, small molecules, metal ions, and even cells. Thus, this assay has significant promise in clinical diagnostics. Sometimes, these are referred to as colorimetric biosensor, which relies on the color change arising from the interparticle plasmon coupling during GNP aggregation (red to purple or blue) or deaggregation (purple to red) [80]. A target analyte or a biological process that directly or indirectly triggers GNP aggregation (or deaggregation) can be detected by the color change of the GNP solution. GNP-based colorimetric assays have high sensitivity, which is comparable to that of conventional fluorescence-based biodetection assays [81]. However, unlike fluorescence-based assays, GNP-based colorimetric/plasmonic assays do not suffer from quenching/destabilization of signal.

4.3.5 Magnetic Biosensing

Magnetic biosensing employs magnetic nanoparticles as proximity sensors that modulate the spin-spin relaxation time of neighboring water molecules, which can be quantified using clinical MRI scanners or benchtop nuclear magnetic resonance (NMR) relaxometers [82, 83]. This technology has been considerably advanced with the development of miniaturized, chip-based NMR detector systems, which are capable of performing highly sensitive measurements on microliter sample volumes and in a multiplexed format. With these and future advances in mind, magnetic biosensor technology promises a high-throughput, low-cost, and portable platform for *in vitro* diagnostics using large-scale molecular and cellular screening.

4.3.6 Electrochemical Biosensing

Electrochemical biosensing is based mainly, but not exclusively, on CNTs. An electrochemical signal from an antibody-antigen interaction on an electrode can be mainly detected using two strategies. The first method detects conductivity changes using CNT in FETs, which was first reported in 1998. Several studies since then have incorporated FETs with immunosensors. The binding of an analyte protein to an antibody attached to the CNT provides a change in electric field, which controls current flow through the device with the condition that the interaction must take place close to the CNT in order to gate the transistor [84, 85].

The second strategy relies on the detection of an analyte with an electroactive tag commonly by using a sandwich assay. In this immunoassay, an antibody is attached to the CNT electrode and the antigen binds to that antibody. The electrode is then exposed to a second antibody that also binds to the antigen. This secondary antibody typically has a tag that is either redox active or can facilitate electrochemiluminescence (ECL) [86].

4.4 *In vivo* Diagnosis, Challenges, and Techniques

While *in vitro* diagnosis is based on simple analysis of isolated/ extracted body fluid samples, *in vivo* diagnosis is a more involved technique as here a number of additional challenges/concerns need to be overcome. These include (a) formulating an injectable sample of the diagnostic probe that maintains its stability in the biological milieu, (b) bypassing the various defense mechanisms/biological barriers of the body that commonly identify, isolate, and degrade any foreign material, (c) reaching the intended target of action amidst the backdrop of several nonspecific sites, and (d) safely excreting out of the body without causing any short- or long-term toxicological effect. For example, from the perspective of pulmonary delivery via the inhalation route, the thick mucosal layer in the lungs constitutes a formidable biological barrier that needs to be overcome in order to gain access to the pulmonary interstitium [6–8]. The fabrication of a nanoprobe that will have all the ideal features needed for targeted delivery *in vivo* is an insurmountable task even for the trained synthetic nanochemist. Nevertheless, a number of diagnostic nanomaterials are being actively investigated for *in vivo* pulmonary diagnosis. The following section will serve as an introduction to various *in vivo* diagnostic techniques, discussing their principal, current limitations and the role of nanotechnology in overcoming these limitations.

4.4.1 Optical Imaging, Including Confocal Endomicroscopy

Optical imaging techniques, which have the highest spatial resolution and sensitivity, can be used for both *in vitro* and *in vivo* studies to determine the interaction of the nanoparticle platform with biological systems [87–89]. High resolution optical imaging methods using fluorescence emission can be used to probe intracellular distribution of molecular events that serve as early signatures of a disease or which indicate a cell response to specific stimuli (e.g., drug response). This multimodal imaging capability of fluorescent nanoparticles can then be specifically targeted to selected pulmonary sites for the assessment of drug efficacy.

A direct application of *in vivo* optical imaging in the lung is provided by confocal fluorescence endomicroscopy of the human airways. This technique aims at providing to the clinician microscopic imaging of a living tissue in real time. The currently available microendoscopic devices use the principle of confocal fluorescent microscopy, in which the objective is replaced by an optical fiber and a miniaturized scanhead at the distal end of the endoscope or by a retractable bundle of optical fibers. Such systems have recently been applied to the endoscopic explorations of several organs, including the gastrointestinal tract, and more recently to the proximal and distal airways *in vivo*. More information about this technique can be found in this review [90].

As in the case with *in vitro* optical imaging, the commonly used optical probes here are based on organic fluorophores, which suffer from poor photostability, thus compromising the detection reliability. For example, methylene blue probes have been used in combination with 660 nm excitation for the cellular imaging of both bronchial epithelial layer and peripheral lung nodules. The replacement of QD-based or other photostable optical nanoprobes can overcome this limitation. In addition, nanoprobes can facilitate multiplexed *in vivo* confocal endoscopy, whereby multiple fluorescence signals can be detected following single-wavelength optical scanning using this technique.

4.4.2 Magnetic Resonance Imaging

MRI is an advanced imaging technique that is able to generate images with high spatial resolution and excellent soft tissue contrast in a noninvasive manner [91–93]. Unlike optical imaging, the MRI signal is independent of the tissue depth and hence can allow deep-tissue imaging. Therefore, this method is widely used in clinical diagnosis and monitoring the response of patients to therapy. MRI relies on the interaction of MR contrast agents with adjacent protons, resulting in strongly influencing their relaxivity. The relaxivities can be either longitudinal (R1, or positive contrast) or transverse (R2, or negative contrast) [93]. The most common contrast agents used for clinical MRI are iron oxide nanoparticles (R1 contrast agents) and gadolinium (Gd) chelates (R2 contrast agents). Although Gd chelates have been the contrast agent of choice for extracellular MRI for more than a decade, the detection sensitivity using these agents is inadequate for molecular MRI [93]. Therefore, in recent years, a significant research effort has been invested toward the fabrication of Gd-containing nanoparticles, which would contain multiple Gd atoms per agent (when compared to Gd chelates with one Gd atom per agent) [26, 35].

4.4.3 Radiographic Imaging

Aerosolized contrast agents promise to improve the resolution of biomedical imaging modalities and enable more accurate diagnosis of lung diseases. Several iodinated compounds, such as diatrizoic acid, have been shown to be safe and useful for radiographic examination of the airways. For example, dry powdered aerosols of diatrizoic acid nanoparticle agglomerates have been demonstrated as a lung contrast agent via radiographic imaging [94]. These nanoparticle agglomerates were created by assembling nanoparticles into inhalable microparticles, which augment deposition in the lung periphery. No acute alveolar tissue damage was observed after two hours of dry powder insufflation to rats.

4.4.4 Multimodal Imaging

There is a growing interest in the development of "fusion technologies," in which two or more medical imaging techniques are combined to provide complementary information. Molecular imaging is another emerging field, in which the modern tools of molecular and cell biology are integrated into advanced imaging tools that allow noninvasive imaging for the early detection of disease onset or real-time monitoring of disease treatment progress. Modern drug discovery is also increasingly reliant on molecular imaging in small animals and even in humans to advance and streamline the drug development process. The use of *in vivo* MRI or radio-imaging enables easier demonstration of drug efficacy, or the lack thereof. This area can also benefit from the development of custom delivery platforms for drugs and imaging contrast agents using molecular markers for targeting.

Nanoparticle-based probes that contain multiple contrast agents for simultaneous use in different imaging modalities (e.g., optical, MR, and radio-imaging) are ideally suited to address this emerging need [1–3, 26, 87]. The correlation of these images using a single nanoprobe will accelerate the diagnostic process while reducing stress on the patient. Another advantage of these "multimodal" nanoprobes is the ability to target specific molecular markers [26, 87]. This capability is increasingly important as biochemical research becomes more successful in characterizing specific molecular factors that accompany a certain disease. These nanoparticle-based contrast agents can be surface modified to carry various biotargeting agents (e.g., antibodies) to target specific disease markers to achieve early detection of diseases. Finally, these nanoprobes have the potential of providing real-time imaging, thus providing the clinician with invaluable information that can be used in designing an individually tailored therapeutic regimen for the patient [87]. Figure 4.1 illustrates a multimodal nanoparticle that combines probes for plasmonic, optical, and MR imaging with the added ability for antibody-conjugated biotargeting.

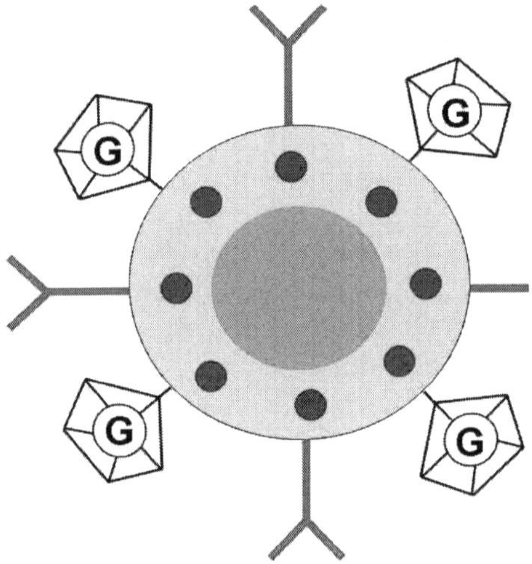

Figure 4.1 The sketch of a multifunctional nanoparticle, where gold nanospheres and QDs/organic fluorophores are coembedded in a silica nanomatrix, with antibodies and chelated gadolinium (G) co-conjugated on their surface. See also Color Insert.

4.5 Specific Examples of Lung Disorders

In principle, an early and accurate diagnosis is critically important for the cure of all diseases. Such diagnosis not only directs a clinician to prescribe the most suitable treatment strategy but also helps in monitoring the therapeutic process in real time. In this section, we will discuss the lung disorders that can immensely benefit from the various *in vitro* and *in vivo* diagnostic techniques, with specific examples of research which has already been or are needed to be conducted.

4.5.1 ALI/ARDS

Acute lung injury (ALI) and its more severe manifestation, acute respiratory distress syndrome (ARDS), occur frequently in trauma patients and have profound personal, medical, societal, and economic impacts [95–97]. Approximately 150,000 cases of ALI and ARDS are diagnosed in the United States each year, often requiring prolonged intensive care and lengthy hospital stays, with associated mortality rates of 25–60%. ALI/ARDS has a long list of etiologies that can be broken down into two categories — 1) direct injury to the lung (e.g., gastric aspiration, pulmonary infections, lung contusion, smoke inhalation, and near-drowning) and 2) indirect injury to the lung (e.g., extrapulmonary sepsis, head injury, pancreatitis, multiple blood transfusions, burns, and uremia). Currently, only supportive care is available for treatment (i.e., supplemental oxygen with low tidal volume, positive-end expiratory pressure mechanical ventilation) with the hope that the patient will eventually recover. Antibiotic treatment is also employed with the assumption of an infectious etiology, though such treatment often has been demonstrated to be counterproductive. There clearly is a need to develop a more specific and selective diagnostic paradigm than is currently available for unwitnessed aspiration events, both those producing frank ALI and subclinical pulmonary injury that may predispose patients to more serious complications, such as ventilator-associated pneumonia (VAP). The development of a diagnostic biomarker assay that can quickly stratify ALI/ARDS patients into the appropriate etiologic category and identify stages of the condition would be a valuable tool to direct appropriate treatment strategies and possibly suggest potential therapy targets. This is particularly important for differentiating gastric aspiration and bacterial pneumonia as they constitute the vast majority of ALI/ARDS cases and present identically in the clinic (at least until pulmonary quantitative bacteriology test results are known).

A number of *in vitro* diagnostic strategies, such as ELISA and flow-cytometry, are being currently used for the quantification of ALI/ARDS biomarkers from body fluids such as the sputum or BAL of suspected patients [62–65]. However, owing to the inadequacies of the assay setup and probes used, these assays suffer from poor sensitivity and are time consuming. In a significant improvement, Knight *et al.* have developed a rapid and sensitive method for simultaneously measuring 16 human protein biomarkers of ALI/ARDS in a microarray-based ELISA technique called MI [73]. Although the sensitivity of this technique is comparable to that of traditional ELISA assays, they too employ organic fluorophores as biological labels, which leads to poor photostability and interassay variability issues, as discussed earlier. The replacement of organic fluorophores with QDs or other photostable nanoparticle-based probes can overcome these interassay variability issues of MI and other assays, thus leading to significant improvements in the diagnosis of ALI/ARDS.

4.5.2 Pneumocystis Pneumonia

Pneumocystis pneumonia (PCP) is a form of pneumonia that afflicts only human beings and is caused by the yeastlike fungus *P. jirovecii*. This pathogen commonly exists in a latent form in the lungs of healthy people. However, it can cause opportunistic lung infection in people with a weak immune system, particularly those using medications that affect the immune system or suffering from cancer, HIV/AIDS, etc. [98]. If untreated, PCP can be a major cause of death in patients suffering from AIDS.

PCP can be diagnosed via rapid and accurate quantification of certain proimflammatory cytokines/protein biomarkers from the sputum or BAL samples of patients. Tasaka *et al.* have characterized the profiles of inflammatory mediators in BAL fluid during PCP in patients with underlying autoimmune diseases, malignancies, or AIDS [98]. Specifically, they measured the concentrations of TNF-alpha, MCP-1, HMGB1, IL-8, IL-6, IL-10, and IFN-gamma in the BAL fluid. They observed that the production of inflammatory mediators in the lung differed between the patient groups with different underlying disorders. However, as in the case of ALI/ARDS diagnosis, rapid and accurate cytokine profiling for PCP is not possible with traditional assay methods as used above. Using QD or GNP-based fluorescence or plasmonic probes, respectively, can enhance the assay sensitivity and reliability of quantifying cytokines for the accurate diagnosis of PCP.

4.5.3 Cystic Fibrosis

Cystic fibrosis (CF), the most common genetic disease among Caucasians, is caused by mutations in the gene encoding cystic fibrosis transmembrane conductance regulator (CFTR) [99, 100]. CF can be diagnosed by measuring the aberrant expression of several biomarkers from *in vitro* samples such as isolated red blood cells and exhaled breath [101]. Fila *et al.* proposed the examination of exhaled breath condensate for the quantification of several types of biomarkers that can indicate CF and other pulmonary diseases [102]. They evaluated the inflammatory acidification of airways and response to antibiotic treatment of pulmonary exacerbation by the examination of the pH of exhaled breath condensate. They also detected from the exhaled breath biomarkers of oxidative stress such as 8-isoprostane and 3-nitrotyrosine, as well as neutrophilic inflammation such as leukotriene B4.

Lange *et al.* have observed that a reduced number of CFTR molecules are expressed in the plasma membrane of erythrocytes in CF patients when compared to that of normal subjects [103]. They used QD-labeled anti-CFTR antibodies as topographic surface markers for their atomic force microscopy (AFM)-mediated detection of individual CFTR molecules. They concluded that erythrocytes reflect the CFTR status of the organism and that quantification of CFTR in a blood sample could be useful in the diagnosis of CF and other related diseases.

Marin *et al.* have detected cystic-fibrosis-related DNA following their linkage with cadmium sulfide QDs and paramagnetic microbeads via direct electrochemical stripping method [104]. In their setup, a cystic-fibrosis-related DNA sequence was sandwiched between two DNA probes, one of which was attached to the paramagnetic microbeads and the other with CdS QDs. The results obtained were successfully employed in a model DNA sensor with an interest in future applications in the clinical field. This developed nanoparticle biosensing system promises significant opportunities in several diseases where rapid, inexpensive, and efficient detection of small volume samples is required.

4.5.4 Tuberculosis

In sharp contrast to cystic fibrosis, which is a major health concern in Western industrialized countries, tuberculosis (TB) is a major killer in developing and third-world countries [105]. This deadly infectious disease is caused by various strains of mycobacteria, usually Mycobacterium TB (MTB) in humans. TB diagnostics is necessary to identify if the infection has progressed to an active disease, or been contained as latent infection, or been eradicated by the host response [106].

Classically, TB is diagnosed based on radiology and growth of the bacterium from collected sputum samples [107]. This diagnostic process is laborious and time consuming. Microscopic analysis of the sputum is also performed prior to culturing; however, the ability to detect very few bacteria in a large sample poses a considerable challenge. Newer diagnostic tests include PCR analysis for specific mycobacterium DNA/RNA markers and interferon γ-release assays from patient cells after exposure to specific mycobacterium proteins [108]. However, these tests require several days and specialized equipment to perform. Antibody-mediated tests for TB are currently becoming available in parts of the world, and there are numerous research programs focused on the development of more specific biomarkers for this disease. However, these methods are time consuming and often unreliable and fail to differentiate between various strains of this mycobacteria.

There is a growing need for a test capable of rapid and sensitive detection of active TB. Rapid diagnosis of pulmonary TB will involve analysis of multiple biomarkers in acquired body fluid samples such as induced sputum, urine, and blood. Several potential biomarkers include (i) interferon γ and tumor necrosis factor α (innate immune response), (ii) specific antibodies to bacterial antigens (active immune response), and (iii) bacterial markers released into blood (ESAT-6, MTB-12, etc.) [109]. A system that has the capability to detect and quantify multiple markers would significantly advance the ability to rapidly diagnose active TB. In addition, these markers can also be utilized in monitoring the efficacy of treatment and resolution of the infection. This can be achieved by using a rapid and multiplexed diagnostic assay using QD-based nanoprobes in high-throughput instrumentation setup, such as a flow-cytometry-based SBA assay or MI, as discussed earlier.

In other nanotechnology-based diagnostic applications, Soo *et al.* have developed a polymerase chain reaction (PCR)-immuno-chromatography test (ICT) for the identification of MTB and differentiation of MTB from other members of M. tuberculosis complex (MTBC) from clinical sputum samples [110, 111]. They further improved the detection flexibility, sensitivity, and cost efficiency by employing an alternative molecular diagnosis assay that used GNPs derivatized with thiol-modified oligonucleotides. The GNP probes, GP-1/GP-2 for IS6110 and GP-3/GP-4 for Rv3618, were designed to specifically hybridize with target DNAs of MTBC and MTB strains, respectively. This assay showed a 96.6% sensitivity and 98.9% specificity toward the detection of MTBC and a 94.7% sensitivity and 99.6% specificity for the detection of MTB.

4.5.5 Lung Cancer

Lung cancer is the most common tumor-related cause of death in Western industrialized countries [112, 113]. The primary reason for the lethality associated with this disease is its improper diagnosis and staging. Conventional diagnostic methods for lung cancer are unsuitable for widespread screening because they are strenuous, expensive, and often unreliable. Improvement of methods for diagnosis of lung cancer, whether *in vitro* or *in vivo*, will help clinicians administer proper therapeutic regimens, thus reversing the dismal picture associated with the prognosis of this disease.

Unlike most of the diseases discussed above, which rely on *in vitro* diagnosis, in lung cancer *in vivo* diagnosis is more critical as tumor diagnosis in its native biological milieu is extremely reliable. In addition to the primary tumor in the lung, this organ is also the most frequent target of metastatic cancer cells originating from other organs. Therefore, various *in vivo* imaging approaches for the diagnosis of both primary tumors and lung metastases are being actively pursued, particularly in "at risk" populations, such as smokers and people with a family history of lung cancer.

Molecular MRI has been used for the sensitive and specific detection of lung metastases, which promises to significantly improve cancer treatment outcomes. Submillimeter-sized metastases with molecular specificity have been detected using high-resolution hyperpolarized (3)He MRI, following targeting of cancer cells with iron oxide nanoparticles functionalized with cancer-binding ligands. The method not only holds promise for cancer imaging but more generally suggests a fundamentally unique approach to molecular imaging in the lungs [114]. Other potential methods of diagnosis of lung cancer *in vivo* involve confocal endomicroscopy (similar to the screening of colon cancer using colonoscopy), radiographic imaging, etc.

Despite the focus on *in vivo* diagnosis in lung cancer, *in vitro* diagnostic methods for the identification of lung-cancer biomarkers or metastatic cancer cells from body-fluid samples deserve sufficient research attention for their simplicity and noninvasive nature. Peng *et al.* have shown that an array of sensors based on GNPs can rapidly distinguish the breath of lung cancer patients from the breath of healthy individuals in an atmosphere of high humidity [115]. In combination with solid-phase microextraction, gas chromatography/mass spectrometry was used to identify 42 volatile organic compounds that represent lung cancer biomarkers. Four of these were used to train and optimize the sensors, demonstrating good agreement between patient and simulated breath samples. Their results show that sensors based on GNPs could form the basis of an inexpensive and noninvasive diagnostic tool for lung cancer.

4.6 Toxicological Studies Using Nanoparticles

Although the ease of drug delivery and the high surface area associated with the lungs can aid in the nanotechnology-mediated diagnosis and therapy of pulmonary diseases, these factors also raise the potential for toxic effects of nanoparticles following their deposition in this organ [17–20, 116]. Pulmonary toxicity of nanoparticles is an area of active research as the human lungs are susceptible to exposure to nanoparticles via several avenues, such as incidental/accidental inhalation from the environment and intentional administration via inhalation or systemic delivery. Although nanotoxicity is not a concern with regard to *in vitro* diagnosis using nanoparticles, the administration of engineered nanoparticles in the lungs for *in vivo* diagnosis should be carefully evaluated in terms of the risks and benefits. Most of these inorganic nanoparticles are nonbiodegradable, and several of them are composed of hazardous elements, such as heavy metals present in QDs. Moreover, inert and presumably benign nanoparticles such as nanoparticulate gold and CNTs are known to exert adverse effects on pulmonary structure and function upon prolonged exposure. The toxic side effects include pulmonary inflammation, fibrosis, carcinogenesis, etc. It is suggested that readers consult some excellent reviews available on this subject [17–20, 116].

4.7 Conclusions

Respiratory and other lung diseases have been and will continue to be a major healthcare concern all over the world. Whether via genetic predisposition, environmental/occupational exposure, accident/trauma, or any other means, human beings are always susceptible to mild or severe pulmonary diseases. Early diagnosis and proper disease stratification/staging will hold the key to successful therapeutic recourse against these diseases. Nanoparticles, owing to their unique properties, will play a critical role in the development of rapid, accurate, ultrasensitive, and "in field" diagnostic methodologies for these diseases. The scientific community is so far skeptical about nanoparticle-mediated *in vivo* diagnosis in the lung owing to potential toxic effects of the nanoparticles. Fortunately, no such skepticism exists concerning nanoparticle-mediated *in vitro* diagnosis from isolated body fluids/cells. Overall, despite concerns, nanotechnology carries tremendous promise in the diagnosis and treatment of pulmonary diseases.

The unique property and potential applications of various nanoparticles are illustrated in Table 4.1, along with related references.

Table 4.1 Table listing the unique properties and potential applications of various nanoparticles

Nanoparticle	Property	Application	References
QDs/rods	Size tunable, photostable, and narrow fluorescence	*In vitro* and *in vivo* optical diagnostics	27–31, 68–71, 88, 89, 103, 104
GNPs/rods	Surface plasmon resonance	*In vitro* colorimetric diagnostics	42–49, 77–81, 111, 115
RE-doped nanocrystals	Tunable, photostable, and narrow fluorescence	*In vitro* and *in vivo* optical diagnostics	32–36
Dye-doped silica/ORMOSIL	Photostable fluorescence, storage stability	*In vitro* and *in vivo* optical diagnostics	21, 38–41

(*Cont'd*)

Table 4.1 (*Cont'd*)

Nanoparticle	Property	Application	References
Iron oxide nanoparticles	Magnetic resonance contrast	*In vitro* and *in vivo* MR diagnostics	50–56, 82, 83, 91, 92, 104
CNTs	Electrical conductivity	*In vitro* electrochemical diagnostics	18, 57–59, 84–86
Diatrizoic acid nanoparticle	Radiographic contrast	*In vivo* radiographic diagnostics	94

Acknowledgments

Sincere thanks go to Dr. Rajiv Kumar for help in preparing this manuscript.

References

1. Farokhzad, O. C., and Langer, R. (2009). Impact of nanotechnology on drug delivery, *ACS Nano*, **3**, pp. 16–20.
2. Whitesides, G. M. (2003). The "right" size in nanobiotechnology, *Nat. Biotechnol.*, **21**, pp. 1161–1165.
3. Pan, D., Lanza, G. M., Wickline, S. A., and Caruthers, S. D. (2009). Nanomedicine: perspective and promises with ligand directed molecular imaging, *Eur. J. Radiol.*, **70**, pp. 274–285.
4. Heath, J. R., and Davis, M. E. (2008). Nanotechnology and cancer, *Annu. Rev. Med.*, **59**, 251–265.
5. Labhasetwar, V. (2005) Nanotechnology for drug and gene therapy: the importance of understanding molecular mechanisms of delivery, *Curr. Opin. Biotechnol.*, **16**, pp. 674–680.
6. Roy, I., and Vij, N. (2010). Nanodelivery in airway diseases: challenges and therapeutic applications, *Nanomedicine*, **6**, pp. 237–244.
7. Hussain, A. A. (1998). Intranasal drug delivery, *Adv. Drug. Del. Rev.*, **29**, pp. 39–49.
8. Yang, W., Peters, J. I., and Williams, III, R. O. (2008). Inhaled nanoparticles — a current review, *Int. J. Pharm.*, **356**, pp. 239–247.
9. Davis, S. S. (1997). Biomedical applications of nanotechnology — implications for drug targeting and gene therapy, *Trends Biotechnol.*, **15**, pp. 217–224.

10. Kommareddy, S., Tiwari, S. B., and Amiji, M. M. (2005). A long-circulating polymeric nanovectors for tumor-selective gene delivery, *Technol. Cancer Res. Treat.*, **4**, 615–625.

11. Rytting, E., Nguyen, J., Wang, X., and Kissel, T. (2008). Biodegradable polymeric nanocarriers for pulmonary drug delivery, *Expert. Opin. Drug. Deliv.*, **5**, pp. 629–639.

12. Lai, S. K., Wang, Y. Y., and Hanes, J. (2009). Mucus-penetrating nanoparticles for drug and gene delivery to mucosal tissues, *Adv. Drug Deliv. Rev.*, **61**, pp. 158–171.

13. Gersting, S. W., Schillinger, U., and Lausier J. (2004). Gene delivery to respiratory epithelial cells by magnetofection, *J. Gene. Med.*, **6**, pp. 913–922.

14. Chakravarthy, K. V., Bonoiu, A. C., Davis, W. G., Ranjan, P., Ding, H., Hu, R., Bowzard, J. B., Bergey, E. J., Katz, J. M., Knight, P. R., Sambhara, S., and Prasad, P. N. (2010). Gold nanorod delivery of an ssRNA immune activator inhibits pandemic H1N1 influenza viral replication, *Proc. Natl. Acad. Sci. USA*, **107**, pp. 10172–10177.

15. Akerman, M. E., Chan, W. C., Laakkonen, P., Bhatia, S. N., and Ruoslahti, E. (2002). Nanocrystal targeting *in vivo*, *Proc. Natl. Acad. Sci. USA*, **99**, pp. 12617–12621.

16. Kurmi, B. D., Kayat, J., Gajbhiye, V., Tekade, R. K., and Jain, N. K. (2010) Micro- and nanocarrier-mediated lung targeting, *Expert. Opin. Drug Deliv.*, **7**, pp. 781–794.

17. Card, J. W., Zeldin, D. C., Bonner, J. C., and Nestmann, E. R. (2008). Pulmonary applications and toxicity of engineered nanoparticles, *Am. J. Physiol. Lung Cell. Mol. Physiol.*, **295**, pp. L400–L411.

18. Shvedova, A. A., and Kagan V. E. (2010). Nanotox-lungs-1 the role of nanotoxicology in realizing the 'helping without harm' paradigm of nanomedicine: lessons from studies of pulmonary effects of single-walled carbon nanotubes, *J. Intern. Med.*, **267**, pp. 106–118.

19. Mansour, H. M., Rhee, Y. S., and Wu, X. (2009). Nanomedicine in pulmonary delivery, *Int. J. Nanomed.*, **4**, pp. 299–319.

20. Schleh, C., and Hohlfeld, J. M. (2009). Interaction of nanoparticles with the pulmonary surfactant system, *Inhal. Toxicol.*, **21**, pp. 97–103.

21. Das, S., Jain, T. K., and Maitra, A. N. (2002). Inorganic-organic hybrid nanoparticles from n-octyl triethoxy silane, *J. Colloids Interface Sci.*, **252**, pp. 82–88.

22. Jain, T. K., Roy, I., De, T. K., and Maitra, A. N. (1998). Nanometer silica particles encapsulating active compounds: a novel ceramic drug carrier, *J. Am. Chem. Soc.*, **120**, pp. 11092–11095.

23. Yokoyoma, M., and Okano, T. (1996). Targetable drug carriers: present status and a future perspective, *Adv. Drug Del. Rev.*, **21**, pp. 77–80.

24. Weetal, H. H. (1970). Storage stability of water insoluble enzymes covalently coupled to organic and inorganic carriers, *Biochim. Biophys. Acta.*, **212**, pp. 1–7.

25. Sharma, P., Brown, S., Walter, G., Santra, S., and Moudgil, B. (2006). Nanoparticles for bioimaging, *Adv. Colloid Interface Sci.*, **123**, pp. 471–485.

26. Prasad, P. N. (2003) *Nanophotonics*, Wiley, New York.

27. Alivisatos, P. (2004). The use of nanocrystals in biological detection, *Nat. Biotechnol.* **22**, pp. 47–52.

28. Choi, H. S., Liu, W., Misra, P., Tanaka, E., Zimmer, J. P., Itty Ipe, B., Bawendi, M. G., and Frangioni, J. V. (2007). Renal clearance of quantum dots, *Nat Biotechnol.*, **25**(10), pp. 1165–1170.

29. Chan, W. C. W., Maxwell, D. J., Gao, X., Bailey, R. E., Hanc, M., and Nie, S. (2002). Luminescent quantum dots for multiplexed biological detection and imaging, *Curr. Opin. Biotechnol.*, **13**, pp. 40–46.

30. Wang, H. Q., Liu, T. C., Cao, Y. C., Huang, Z. L., Wang, J. H., Li, X. Q., and Zhao, Y. D. (2006). A flow cytometric assay technology based on quantum dots-encoded beads, *Anal. Chim. Acta.*, **580**, pp. 18–23.

31. Liu, T., Liu, B., Zhang, H., and Wang, Y. (2005). The fluorescence bioassay platforms on quantum dots nanoparticles, *J. Fluoresc.*, **15**, pp. 729–733.

32. Karar, N., and Chander, H. (2005). Nanocrystal formation and luminescence properties of ZnS based doped nanophosphors, *J. Nanosci. Nanotechnol.*, **5**, pp. 1498–1502.

33. Lim, S. F., Riehn, R., Ryu, W. S., Khanarian, N., Tung, C. K., Tank, D., and Austin, R. H. (2006). *In vivo* and scanning electron microscopy imaging of up-converting nanophosphors in Caenorhabditis elegans, *Nano Lett.*, **6**, pp. 169–174.

34. Nyk, M., Kumar, R., Ohulchanskyy, T. Y., Bergey, E. J., and Prasad, P. N. (2008). High contrast *in vitro* and *in vivo* photoluminescence bioimaging using near infrared to near infrared up-conversion in Tm^{3+} and Yb^{3+} doped fluoride nanophosphors, *Nano Lett.*, **8**, pp. 3834–3838.

35. Kumar, R., Nyk, M., Ohulchanskyy, T. Y., Flask, C. A., and Prasad, P. N. (2009). Combined optical and MR bioimaging using rare earth ion doped $NaYF_4$ nanocrystals, *Adv. Func. Mater.*, **19**, pp. 853–859.

36. Chen, X. Y., Ma, E., and Liu, G. K. (2007). Energy levels and optical spectroscopy of Er^{3+} in Gd_2O_3 nanocrystals, *J. Phys. Chem. C*, **111**, pp. 10404–10411.

37. Buck, S. M., Xu, H., Brasuel, M., Philbert, M. A., and Kopelman, R. (2004). Nanoscale probes encapsulated by biologically localized embedding (PEBBLEs) for ion sensing and imaging in live cells, *Talanta*, **63**, pp. 41–59.

38. Sokolov, I., and Naik, S. (2008). Novel fluorescent silica nanoparticles: towards ultrabright silica nanoparticles, *Small*, **4**, pp. 934–939.

39. Xing, X. L., He, X. X., Peng, J. F., Wang, K. M., and Tan, W. H. (2005). Uptake of silica-coated nanoparticles by HeLa cells, *J. Nanosci. Nanotechnol.*, **5**, pp. 1688–1693.

40. Sharma, R. K., Das, S., and Maitra, A. (2004). Surface modified ormosil nanoparticles, *J. Colloid Interface Sci.*, **277**, pp. 342–346.

41. Roy, I., Ohulchanskyy, T. Y., Pudavar, H. E., Bergey, E. J., Oseroff, A. R., Morgan, J., Dougherty, T. J., and Prasad, P. N. (2003). Ceramic-based nanoparticles entrapping water-insoluble photosensitizing anticancer drugs: a novel drug-carrier system for photodynamic therapy, *J. Am. Chem. Soc.*, **125**, pp. 7860–7865.

42. Huang, X., Jain, P. K., El-Sayed, I. H., and El-Sayed, M. A. (2007) Gold nanoparticles: Interesting optical properties and recent applications in cancer diagnostics and therapy, *Nanomed.*, **2**, pp. 681–693.

43. Hu, R., Yong, K. T., Roy, I., Ding, H., He, S., and Prasad, P. N. (2009). Metallic nanostructures as localized plasmon resonance enhanced scattering probes for multiplex dark-field targeted imaging of cancer cells, *J. Phys. Chem. C*, **113**, pp. 2676–2684.

44. Uechi, I., and Yamada, S. (2008). Photochemical and analytical applications of gold nanoparticles and nanorods utilizing surface plasmon resonance, *Anal. Bioanal. Chem.*, **391**, pp. 2411–2421.

45. Arvizo, R., Bhattacharya, R., and Mukherjee, P. (2010). Gold nanoparticles: opportunities and challenges in nanomedicine, *Expert Opin. Drug Deliv.*, **7**, pp. 753–763.

46. Ahmad, M. Z., Akhter, S., Jain, G. K., Rahman, M., Pathan, S. A., Ahmad, F. J., and Khar, R. K. (2010). Metallic nanoparticles: technology overview & drug delivery applications in oncology, *Expert Opin. Drug Deliv.*, **7**, pp. 927–942.

47. Ghosh, P., Han, G., De, M., Kim, C. K., and Rotello, V. M. (2008). Gold nanoparticles in delivery applications, *Adv. Drug Deliv. Rev.*, **60**, pp. 1307–1315.

48. Patra, C. R., Bhattacharya, R., Mukhopadhyay, D., and Mukherjee, P. (2010). Fabrication of gold nanoparticles for targeted therapy in pancreatic cancer, *Adv. Drug Deliv. Rev.*, **62**, pp. 346–361
49. Giordano, R. J., Edwards, J. K., Tuder, R. M., Arap, W., and Pasqualini, R. (2009). Combinatorial ligand directed lung targeting, *Proc. Am. Thorac. Soc.*, **6** pp. 411–415.
50. Burtea, C., Laurent, S., Vander, Elst L., and Muller, R. N. (2008). Contrast agents: magnetic resonance, *Handb. Exp. Pharmacol.*, **185 Pt 1**, pp. 135–165.
51. Frullano, L., and Meade, T. J. (2007). Multimodal MRI contrast agents, *J. Biol. Inorg. Chem.*, **12**, pp. 939–949.
52. Lanza, G. M., Winter, P. M., Caruthers, S. D., Morawski, A. M., Schmieder, A. H., Crowder, K. C., and Wickline, S. A. (2004). Magnetic resonance molecular imaging with nanoparticles, *J. Nucl. Cardiol.*, **11**, pp. 733–743.
53. Petersein, J., Saini, S., and Weissleder, R. (1996). Liver. II: Iron oxide-based reticuloendothelial contrast agents for MR imaging, *Magn. Reson. Imaging Clin. N Am.*, **4**, pp. 53–60.
54. Wittenberg, N. J., and Haynes, C. L. (2009). Using nanoparticles to push the limits of detection, *Wiley Interdiscip. Rev. Nanomed. Nanobiotechnol.*, **1**, pp. 237–254.
55. Palecek, E., and Fojta, M. (2007). Magnetic biosensor magnetic beads as versatile tools for electrochemical DNA and protein biosensing, *Talanta*, **74**, pp. 276–290.
56. Haun, J. B., Yoon, T. J., Lee, H., and Weissleder, R. (2010). Magnetic nanoparticle biosensors, *Wiley Interdiscip. Rev. Nanomed. Nanobiotechnol.*, **2**, pp. 291–304.
57. Tenne, R., and Rao, C. N. (2004). Inorganic nanotubes, *Philos. Transact. A Math Phys. Eng. Sci.*, **362**, pp. 2099–2125.
58. Hanus, M. J., and Harris, A. T. (2010). Synthesis, characterisation and applications of coiled carbon nanotubes, *J. Nanosci. Nanotechnol.*, **10**, pp. 2261–2283.
59. Myung, S., Woo, S., Im, J., Lee, H., Min, Y. S., Kwon, Y. K., and Hong S. (2010) Large-scale assembly of 'type-switchable' field effect transistors based on carbon nanotubes and nanoparticles, *Nanotechnology*, **21**, pp. 345301–354305.
60. Remick, D. G., Bolgos, G. R., Siddiqui, J., Shin, J., and Nemzek, J. A. (2002). Six at six: interleukin-6 measured 6 h after the initiation of sepsis predicts mortality over 3 days, *Shock*, **17**, pp. 463–467.

61. Van der Poll, T., and Lowry, S. F. (1993). Tumor necrosis factor in sepsis: mediator of multiple organ failure or essential part of host defense?, *Shock*, **3**, pp. 1–12.

62. Nemzek, J. A., Siddiqui, J., and Remick, D. G. (2001). Development and optimization of cytokine ELISAs using commercial antibody pairs, *J. Immunol. Methods*, **255**, pp. 149–157.

63. Elshal, M. F., and McCoy, J. P. (2006). Multiplex bead array assays: performance evaluation and comparison of sensitivity to ELISA, *Methods*, **38**, pp. 317–323.

64. Lund-Johansen, F., Davis, K., Bishop, J., and de Waal Malefyt, R. (2000). Flow cytometric analysis of immunoprecipitates: high-throughput analysis of protein phosphorylation and protein-protein interactions, *Cytometry*, **39**, pp. 250–259.

65. Hsu, H. Y., Wittemann, S., Schneider, E. M., Weiss, M., and Joos, T. O. (2008). Suspension microarrays for the identification of the response patterns in hyperinflammatory diseases, *Med. Eng. Phys.*, **30**, pp. 976–983.

66. Heijmans-Antonissen, C., Wesseldijk, F., Munnikes, R. J., Huygen, F. J., van der Meijden, P., Hop, W. C., Hooijkaas, H., and Zijlstra, F. J. (2006). Multiplex bead array assay for detection of 25 soluble cytokines in blister fluid of patients with complex regional pain syndrome type 1, *Mediators Inflamm.*, **1**, pp. 28398–28405.

67. Khan, S. S., Smith, M. S., Reda, D., Suffredini, A. F., and McCoy, J. P., Jr. (2004). Multiplex bead array assays for detection of soluble cytokines: comparisons of sensitivity and quantitative values among kits from multiple manufacturers, *Cytometry B Clin. Cytom.*, **61**, pp. 35–39.

68. Liu, T., Liu, B., Zhang, H., and Wang, Y. (2005). The fluorescence bioassay platforms on quantum dots nanoparticles, *J. Fluoresc.*, **15**, 729–733.

69. Abrams, B., and Dubrovsky, T. (2007). Quantum dots in flow cytometry, *Methods Mol. Biol.*, **374**, pp. 185–203.

70. Chattopadhyay, P. K., Yu, J., and Roederer, M. (2007). Application of quantum dots to multicolor flow cytometry, *Methods Mol. Biol.*, **374**, pp. 175–184.

71. Ferrari, B. C., and Bergquist, P. L. (2007). Quantum dots as alternatives to organic fluorophores for Cryptosporidium detection using conventional flow cytometry and specific monoclonal antibodies: lessons learned, *Cytometry A*, **71**, pp. 265–271.

72. Leng, S. X., McElhaney, J. E., Walston, J. D., Xie, D., Fedarko, N. S., and Kuchel, G. A. (2008). ELISA and multiplex technologies for cytokine

measurement in inflammation and aging research, *J. Gerontol. A Biol. Sci. Med. Sci.*, **63**, pp. 879–884.

73. Knight, P. R., Sreekumar, A., Siddiqui, J., Laxman, B., Copeland, S., Chinnaiyan, A., and Remick, D. G. (2004). Development of a sensitive microarray immunoassay and comparison with standard enzyme-linked immunoassay for cytokine analysis, *Shock*, **21**, pp. 26–30.
74. Li, J., Cao, Z. C., Tang, Z., Wang, K., and Tan, W. (2008). Molecular beacons for protein-DNA interaction studies, *Methods Mol. Biol.*, **429**, pp. 209–224.
75. Silverman, A. P., and Kool, E. T. (2005). Quenched probes for highly specific detection of cellular RNAs, *Trends Biotechnol.*, **23**, pp. 225–230.
76. Santangelo, P. J. (2010). Molecular beacons and related probes for intracellular RNA imaging, *Wiley Interdiscip. Rev. Nanomed. Nanobiotechnol.*, **2**, pp. 11–19.
77. Mao, X., Xu, H., Zeng, Q., Zeng, L., and Liu, G. (2009). Molecular beacon-functionalized gold nanoparticles as probes in dry-reagent strip biosensor for DNA analysis, *Chem. Commun. (Camb).*, **7**, pp. 3065–3067.
78. Anker, J. N., Hall, W. P., Lyandres, O., Shah, N. C., Zhao, J., and Van Duyne, R. P. (2008). Biosensing with plasmonic nanosensors, *Nat. Mater.*, **7**, pp. 442–453.
79. Vo-Dinh, T., Wang, H. N., and Scaffidi, J. (2010). Plasmonic nanoprobes for SERS biosensing and bioimaging, *J Biophotonics.*, **3**, pp. 89–102.
80. Zhao, W., Brook, M. A., and Li, Y. (2008). Design of gold nanoparticle-based colorimetric biosensing assays, *ChemBioChem*, **9**, pp. 2363–2371.
81. Stuart, D. A., Haes, A. J., Yonzon, C. R., Hicks, E. M., and Van Duyne, R. P. (2005). Biological applications of localised surface plasmonic phenomenae, *IEE Proc. Nanobiotechnol.*, **152**, pp. 13–32.
82. Danielli, A., Porat, N., Ehrlich, M., and Arie, A. (2009). Magnetic modulation biosensing for rapid and homogeneous detection of biological targets at low concentrations, *Curr. Pharm. Biotechnol.*, **11**, pp. 128–137.
83. Palecek, E., and Fojta, M. (2007). Magnetic beads as versatile tools for electrochemical DNA and protein biosensing, *Talanta*, **74**, pp. 276–290.
84. Wang, J., and Lin, Y. (2008). Functionalized carbon nanotubes and nanofibers for biosensing applications, *Trends Analyt. Chem.*, **27**, pp. 619–626.
85. Pumera, M. (2009). The electrochemistry of carbon nanotubes: fundamentals and applications, *Chemistry*, **15**, pp. 4970–4978.

86. Jacobs, C. B., Peairs, M. J., and Venton, B. J. (2010). Review: carbon nanotube based electrochemical sensors for biomolecules, *Anal. Chim. Acta.*, **662**, pp. 105–127.

87. Prasad, P. N. (2004) *Introduction to Biophotonics*, Wiley, New York.

88. Tholouli, E., Sweeney, E., Barrow, E., Clay, V., Hoyland, J. A., and Byers, R. J. (2008). Quantum dots light up pathology, *J. Pathol.*, **216**, pp. 275–285.

89. Santra, S., Xu, J., Wang, K., and Tan, W. (2004). Luminescent nanoparticle probes for bioimaging, *J. Nanosci. Nanotechnol.*, **4**, pp. 590–599.

90. Thiberville, L., Salaün, M., Lachkar, S., Dominique, S., Moreno-Swirc, S., Vever-Bizet, C., and Bourg-Heckly, G. (2009). Confocal fluorescence endomicroscopy of the human airways, *Proc. Am. Thorac. Soc.*, **6**, pp. 444–449.

91. Beckmann, N., Cannet, C., Babin, A. L., Blé, F. X., Zurbruegg, S., Kneuer, R., and Dousset, V. (2009). *In vivo* visualization of macrophage infiltration and activity in inflammation using magnetic resonance imaging, *Wiley Interdiscip. Rev. Nanomed. Nanobiotechnol.*, **1**, pp. 272–298.

92. Peng, X. H., Qian, X., Mao, H., Wang, A. Y., Chen, Z. G., Nie, S., and Shin, D. M. (2008). Targeted magnetic iron oxide nanoparticles for tumor imaging and therapy, *Int. J. Nanomedicine.*, **3**, pp. 311–321.

93. Sosnovik, D. E., Nahrendorf, M., and Weissleder, R. (2007). Molecular magnetic resonance imaging in cardiovascular medicine, *Circulation* **115**, pp. 2076–2086.

94. El-Gendy, N., Aillon, K. L., and Berkland, C. (2010). Dry powdered aerosols of diatrizoic acid nanoparticle agglomerates as a lung contrast agent, *Int. J. Pharm.*, **391**, pp. 305–312.

95. ARDS Network. (2000). Ketoconazole for early treatment of acute lung injury and acute respiratory distress syndrome: a randomized controlled trial, *The ARDS Network JAMA*, **283**, pp. 1995–2002.

96. Bernard, G. R., Artigas, A., Brigham, K. L., Carlet, J., Falke, K., Hudson, L., Lamy, M., Legall, J. R., Morris, A., and Spragg, R. (1994). The American-European Consensus Conference on ARDS: definitions, mechanisms, relevant outcomes, and clinical trial coordination, *Am. J. Respir. Crit. Care. Med.*, **149**, pp. 818–824.

97. Hirvela, E. R. (2000). Advances in the management of acute respiratory distress syndrome: protective ventilation, *Arch. Surg.*, **135**, pp. 126–135.

98. Tasaka, S., Kobayashi, S., Kamata, H., Kimizuka, Y., Fujiwara., H., Funatsu, Y., Mizoguchi, K., Ishii, M., Takeuchi, T., and Hasegawa, N. (2010). Cytokine profiles of bronchoalveolar lavage fluid in patients with pneumocystis pneumonia, *Microbiol. Immunol.*, **54**, pp. 425–433.

99. Rowe, S. M., Miller, S., and Sorscher E. J. (2005). Cystic fibrosis, *N. Engl. J. Med.*, **352**, pp. 1992–2001.
100. Wine, J. J. (1999). The genesis of cystic fibrosis lung disease, *J. Clin. Invest.*, **103**, pp. 309–312.
101. Belcher, C. N., and Vij, N. (2010). Protein processing and inflammatory signaling in cystic fibrosis: challenges and therapeutic strategies, *Curr. Mol. Med.*, **10**, pp. 82–94.
102. Fila, L., and Musil, J. (2010). Examination of exhaled breath condensate in cystic fibrosis, *Cas. Lek. Cesk.*, **149**, pp. 173–177.
103. Lange, T., Jungmann, P., Haberle, J., Falk, S., Duebbers, A., Bruns, R., Ebner, A., Hinterdorfer, P., Oberleithner, H., and Schillers, H. (2006). Reduced number of CFTR molecules in erythrocyte plasma membrane of cystic fibrosis patients, *Mol. Membr. Biol.*, **23**, pp. 317–323.
104. Marin, S., and Merkoçi, A. (2009). Direct electrochemical stripping detection of cystic-fibrosis-related DNA linked through cadmium sulfide quantum dots, *Nanotechnology*, **20**, pp. 1–6.
105. Courtwright, A., and Turner, A. N. (2010). Tuberculosis and stigmatization: pathways and interventions, *Public Health Rep.*, **125**, pp. 34–42.
106. Wallis, R. S., Pai, M., Menzies, D., Doherty, T. M., Walzl, G., Perkins, M. D., and Zumla, A. (2010). Biomarkers and diagnostics for tuberculosis: progress, needs, and translation into practice, *Lancet*, **375**, pp. 1920–1937.
107. Horne, D. J., Royce, S. E., Gooze, L., Narita, M., Hopewell, P. C., Nahid, P., and Steingart K. R. (2010). Sputum monitoring during tuberculosis treatment for predicting outcome: systematic review and meta-analysis, *Lancet Infect. Dis.*, **10**, pp. 387–394
108. Lange, C., and Mori, T. (2010). Advances in the diagnosis of tuberculosis, *Respirology*, **15**, pp. 220–240.
109. Horne, D. J., Royce, S. E., Gooze, L., Narita, M., Hopewell, P. C., Nahid, P., and Steingart K. R. (2010). Sputum monitoring during tuberculosis treatment for predicting outcome: systematic review and meta-analysis, *Lancet Infect. Dis.*, **10**, pp. 387–394
110. Soo, P. C., Horng, Y. T., Hsueh, P. R., Shen, B. J., Wang, J. Y., Tu, H. H., Wei, J. R., Hsieh, S. C., Huang, C. C., and Lai, H. C. (2006). Direct and simultaneous identification of Mycobacterium tuberculosis complex (MTBC) and Mycobacterium tuberculosis (MTB) by rapid multiplex nested PCR-ICT assay, *J. Microbiol. Methods*, **66**, pp. 440–448.

111. Soo, P. C., Horng, Y. T., Chang, K. C., Wang, J. Y., Hsueh, P. R., Chuang, C. Y., Lu, C. C., and Lai, H. C. (2009). A simple gold nanoparticle probes assay for identification of Mycobacterium tuberculosis and Mycobacterium tuberculosis complex from clinical specimens, *Mol. Cell Probes.*, **23**, pp. 240–246.

112. Weissferdt, A., and Moran, C. A. (2010). Primary vascular tumors of the lungs: a review, *Ann. Diagn. Pathol.* **14**, pp. 296–308.

113. Stinchcombe, T. E., Bogart, J., Wigle, D. A., and Govindan, R. (2010). Annual review of advances in lung cancer clinical research: a report for the year 2009, *J. Thorac. Oncol.*, **5**, pp. 935–939.

114. Branca, R. T., Cleveland, Z. I., Fubara, B., Kumar, C. S., Maronpot, R. R., Leuschner, C., Warren, W. S., and Driehuys, B. (2010). Molecular MRI for sensitive and specific detection of lung metastases, *Proc. Natl. Acad. Sci. USA*, **107**, pp. 3693–3697.

115. Peng, G., Tisch, U., Adams, O., Hakim, M., Shehada, N., Broza, Y. Y., Billan, S., Abdah-Bortnyak, R., Kuten, A., and Haick, H. (2009). Diagnosing lung cancer in exhaled breath using gold nanoparticles, *Nat. Nanotechnol.* **4**, pp. 669–673.

116. Buxton, D. B. (2011) Nanomedicine for the management of lung and blood diseases, *Nanomedicine (Lond)*, **4**, pp. 331–339.

Chapter 5

Nanoparticles for Targeting T Cells in Allergy and Inflammatory Airway Conditions

Adham Bear,[a] Laura B. Carpin,[b] Conrad R. Cruz,[a] Rebekah A. Drezek,[b] and Aaron E. Foster[a,*]

[a]*Center for Cell and Gene Therapy, Baylor College of Medicine, Texas Children's Hospital and the Methodist Hospital, Houston, TX 77030, USA*
[b]*Department of Bioengineering, Rice University Houston, TX 77005, USA*
[*]aaronf@bcm.edu

5.1 Introduction

T cells have been implicated in the pathogenesis of various chronic inflammatory airway conditions such as asthma and chronic obstructive pulmonary disease. The correlation between T cells and disease is the strongest in the case of allergic asthma, where disease initiation and progression can be attributed to a single T cell subset. Asthma is commonly treated with inhaled β2 agonists and inhaled corticosteroids and for exacerbations, oral corticosteroids. Although effective, these drugs broadly suppress inflammatory and immune responses and have the risk of serious side effects. Nanotechnology presents the opportunity to deliver a concentrated and localized dose of therapeutic, reducing or eliminating the risk of side effects. In addition, nanoparticles can be modified to permit cell-specific targeting, which could be valuable in the treatment of

Pulmonary Nanomedicine: Diagnostics, Imaging, and Therapeutics
Edited by Neeraj Vij

ISBN 978-981-4316-48-4 (Hardback), 978-981-4364-14-0 (eBook)
www.panstanford.com

asthma. Therefore, this chapter will focus on targeting T cells with therapeutic nanoparticles in the context of asthma treatment. We will begin with a discussion of the role of T cells in the pathogenesis of asthma. Next, we will briefly discuss current strategies for the treatment of asthma and how T cell–directed nanoparticles may help to overcome some of the limitations of current treatments. We will then discuss the T cell structure in detail to highlight potential surface targets to which nanoparticles may be directed. Finally, we will conclude with a discussion of current targeting ligands and provide examples of how they have been used in previous studies to target T cells both *in vitro* and *in vivo*.

5.2 Role of T Cells in the Pathogenesis of Asthma

T cell involvement in inflammatory airway conditions is perhaps most well characterized in allergic asthma [Herrick *et al.*, 2003]. Asthma is a chronic inflammatory lung condition characterized by airway hyperreactivity and mucus hypersecretion. The clinical syndrome of asthma can be attributed to various etiologies and displays a variety of histopathologic properties. However, the bronchial mucosa of atopic asthmatics is unique due to characteristic lymphocyte and eosinophil infiltration. Other findings include goblet-cell hyperplasia and mucus plugging, bronchial smooth muscle hyperplasia and hypertrophy, subbasement membrane thickening and collagen deposition, and mast cell degranulation [Busse *et al.*, 2001]. All of these findings work in concert to cause intermittent airway obstruction following allergen exposure.

T cells were first implicated in the pathogenesis of asthma following the association of asthma severity with immunoglobulin E (IgE) antibodies [Burrows *et al.*, 1989]. The initiation of IgE synthesis requires an adaptive immune response contingent on a specific T cell subset (Fig. 5.1). Upon inhalation of allergens, dendritic cells (DCs) lining the airway engulf and process antigens. DCs then migrate to draining lymph nodes, where they encounter T cells and B cells. Depending on the context in which DCs present antigen to T cells, a $CD4^+$ T Helper 1 or 2 (T_H1 or T_H2) response is generated [Kim *et al.*, 1985; Mosmann *et al.*, 1986]. Each subset can be distinguished by a unique cytokine profile and has distinct effector functions.

T_H1 cells secrete inflammatory cytokines like interferon-γ (IFN-γ) and tumor necrosis factor-α (TNF-α), which activate macrophages for the clearance of intracellular pathogens. Conversely, T_H2 cells secrete interleukin (IL)-4, IL-5, IL-9, and IL-13 which promote antibody responses from B cells and activate eosinophils and mast cells for the clearance of parasites and helminths. It is the T_H2 cell that directs B cell IgE class switching through the secretion of IL-4 and IL-13 as well as through CD40-CD40L cell surface interactions [Bacharier *et al.*, 1998]. Therefore, asthma is the direct result of T_H2-mediated IgE secretion by B cells that results in the triggering of an innate immune response.

Indeed, high levels of T_H2 cytokines have been detected in the bronchial alveolar lavage fluid of atopic asthmatics [Robinson *et al.*, 1993; Robinson *et al.*, 1992; Walker *et al.*, 1992], and asthma can be induced in mice by the adoptive transfer of T-cell receptor transgenic T_H2 cells [Cohn *et al.*, 1998]. Furthermore, the induction of asthma can be prevented by the *in vivo* depletion of $CD4^+$ T cells [Gavett *et al.*, 1994] or in the absence of IL-4 [Brusselle *et al.*, 1994; Coyle *et al.*, 1995], which is necessary for the priming of T_H2 cells. However, once T cell priming occurs, airway inflammation occurs in an IL-4-independent manner [Corry *et al.*, 1996; Coyle *et al.*, 1995]

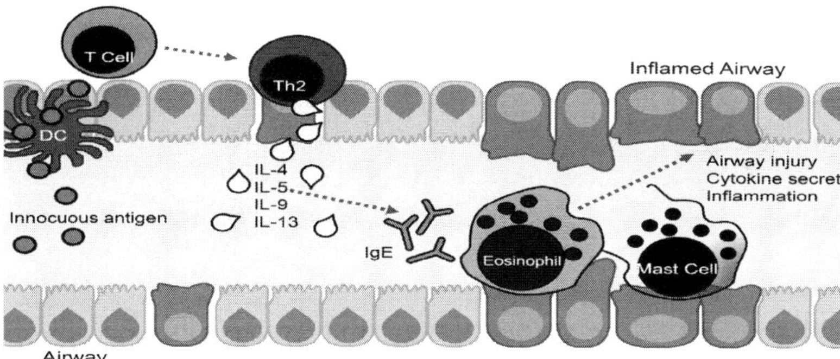

Figure 5.1 The role of T_H2 cells in the initiation of atopic asthma. DCs lining the airway process antigens and prime T_H2-type responses within lymph nodes. T_H2 cells secrete IL-4, IL-5, IL-9, and IL-13 upon encountering antigen within the airway. These cytokines promote B cell IgE class switching. IgE secretion by B cells then triggers innate immune responses mediated by mast cells and eosinophils, leading to airway inflammation. See also Color Insert.

and cytokines such as IL-5 and IL-13 play more important roles, causing eosinophil infiltration and mucus hypersecretion [Foster *et al.*, 1996; Grunig *et al.*, 1998; Kung *et al.*, 1995; Lee *et al.*, 1997; Van Oosterhout *et al.*, 1995; Walter *et al.*, 2001; Wills-Karp *et al.*, 1998; Zhu *et al.*, 1999].

Recent advances in the field of immunology have suggested a potential role of other T cell subtypes in the pathogenesis of asthma [Robinson, 2009]. T_H17 cells are a recently described subset of $CD4^+$ T cells. T_H17 cells are characterized by the production of IL-17 and have been implicated in a variety of autoimmune diseases [Langrish *et al.*, 2005]. IL-17 expression by T cells has been detected in the airways of asthmatics [Molet *et al.*, 2001], and the cotransfer of T_H17 cells along with T_H2 cells augments airway inflammation in mouse models [Wakashin *et al.*, 2008]. On the contrary, regulatory T cells (T_{reg}) may play a role in decreasing T cell responses to allergens. T_{reg} are a subset of $CD4^+$ T cells characterized by high expression of CD25, the IL-2 receptor α chain. Unlike other T_H subsets, T_{reg} do not proliferate or produce cytokines following antigenic stimulation but rather suppress proliferation and cytokine production by effector T cells. Suppression is carried out via the expression of inhibitory cytokines, such as IL-10 and transforming growth factor-β (TGF-β), and the expression of inhibitory molecules, such as cytotoxic lymphocyte antigen-4 (CTLA-4) and programmed death-1 (PD-1). The depletion of $CD4^+CD25^+$ T cells from nonatopic individuals results in the proliferation of and cytokine production by $CD4^+CD25^-$ T cells *in vitro* [Ling *et al.*, 2004], and IL-10-producing T cells can suppress T_H2 cytokine production [Akdis *et al.*, 2004]. Furthermore, mouse studies have demonstrated that the transfer of antigen-specific T_{reg} cells reduces airway hyperreactivity, eosinophil recruitment, and T_H2 cytokine production in an IL-10-dependent manner [Kearley *et al.*, 2005].

Having a basic understanding of T cell involvement in asthma pathogenesis is essential when considering potential therapeutic targets. Nanoparticles could be invaluable in future asthma treatments by providing a mechanism to target the specific subsets of T cells that are related to the pathogenesis of asthma. In the following section, we will discuss the current therapeutic strategies employed that broadly suppress inflammatory responses. We will also explore two potential nanoparticle-based treatment strategies for inhibiting the T_H2 allergic cascade — (1) direct inhibition of instigating T_H2 cells using steroids and (2) shifting of the T_H1:T_H2 balance by inducing T_H1-type responses.

5.3 Treatment Strategies for Asthma

5.3.1 Nanosteroids for the Treatment of Asthma

T_H2 cells act as orchestrators of the allergic response in asthma, secreting cytokines that induce B cell class switching to IgE synthesis (IL-4 and IL-13), mast cell recruitment (IL-4, IL-9, and IL-13), and eosinophil maturation (IL-3, IL-5, and GM-CSF). The expression of cytokines by T cells is coordinated by several transcription factors, such as nuclear factor-κB (NF-κB), activator protein-1 (AP-1), nuclear factor of activated T cells (NF-AT), cyclic AMP response element binding protein (CREB), and signal transduction-activated transcription factors (STAT) [Barnes *et al.*, 1998]. Therefore, the inhibition of these transcription factors within this specific cell population may block cytokine secretion and prevent the downstream effector cells of the allergic response (Fig. 5.2).

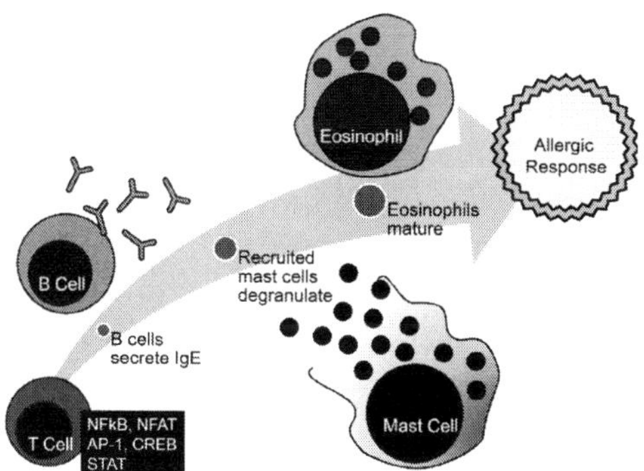

Figure 5.2 Transcription factor involvement in the pathogenesis of asthma. The transcription factors NF-κB, NF-AT, AP-1, CREB, and STAT coordinate T cell cytokine secretion. Cytokine secretion by T_H2 cells causes a cascade of B cell IgE class switching, mast cell degranulation, and eosinophil maturation, leading to an allergic response. Corticosteroids inhibit T cell cytokine secretion by interfering with transcription factor function, thereby inhibiting the downstream effector cell function and decreasing the allergic response. See also Color Insert.

Inhaled or oral glucocorticoids are the most effective treatment for the long-term control of asthma because they both stimulate the expression of genes encoding anti-inflammatory mediators and inhibit the expression of inflammatory genes. This dual effect is carried out by two distinct processes — transactivation and transrepression [Barnes *et al.*, 2003]. Corticosteroids are highly lipophilic molecules that easily permeate cell membranes and subsequently bind to cytoplasmic glucocorticoid receptors (GR). The GR-corticosteroid complex then translocates to the nucleus where gene regulation occurs. Through transactivation, GR-corticosteroid complexes bind to glucocorticoid response elements (GRE) in the promoter regions of steroid-responsive genes, leading to transcriptional activation of various anti-inflammatory genes, such as annexin-1, IL-10, and IκBα (the inhibitor of NF-κB). Through transrepression, GR-corticosteroid complexes bind large coactivator molecules that are activated by transcription factors like NF-κB and AP-1 and thereby inhibit the expression of inflammatory genes [Holgate *et al.*, 2008].

Despite the proven efficacy of glucocorticoids for the treatment of asthma, major concerns regarding their specificity exist. Glucocorticoid receptors are expressed on most cell types in the body, and treatment may result in severe systemic side effects, which differ in character and severity based on the route of administration. Complications of inhaled glucocorticoids are usually less severe and include thrush, osteoporosis, adrenal suppression, and cataracts. In comparison, the complications associated with systemic corticosteroids may be more pronounced, often leading to immune suppression, diabetes, and cardiovascular disease [Barnes, 1995; Dahl, 2006].

One way to avoid the systemic side effects involved with glucocorticoid therapy is to target cell-specific transcription factors. One example is NF-AT. NF-AT is a transcription factor predominantly expressed by T lymphocytes and regulates the transcription of cytokines like IL-2, IL-4, and IL-5 [Hodge *et al.*, 1995; Hodge *et al.*, 1996; Lee *et al.*, 1994]. Drugs like cyclosporin A and tacrolimus (FK506), which block NF-AT activation by inhibiting calcineurin, provide a means of selectively targeting T cells and inhibiting T_H2 responses. In fact, cyclosporin A has demonstrated efficacy as a treatment for glucocorticoid-resistant asthmatics [Alexander *et al.*, 1992], demonstrating that T cell targeted therapies may have therapeutic benefits for the treatment of asthma. However, cyclosporine A and tacrolimus treatment can also have serious systemic side effects, which include nephrotoxicity, hypertension, neurotoxicity, and complications of immune suppression, such as increased risk of infection or malignancy. Novel drug delivery systems incorporating new technologies are being developed to enhance the localization and sustained release of drugs at the target sites. Such systems may allow for the administration of lower drug doses as well as reduce drug delivery to off-target sites, thereby decreasing the frequency and severity of systemic side effects.

Nanoparticles have emerged as a new class of drug delivery vehicles that permit a variety of drug encapsulation techniques [Peer *et al.*, 2007]. Ishihara *et al.* [2005] developed and characterized betamethasone-encapsulated poly(lactic-co-glycolic acid) (PLGA) nanosteroid particles. The authors demonstrated that by using PLGA, a biodegradable polymer, they could achieve over eight days of sustained steroid release from macrophages that had engulfed particles *in vitro*. Higaki *et al.* [2005] then demonstrated that PLGA nanosteroids administered intravenously efficiently reduced inflammation in both rat and mouse models of inflammatory arthritis to a greater extent than the free drug when administered alone. Furthermore, intravenously administered PLGA nanosteroids were shown to localize to inflamed sites and reduce the severity of inflammation in a mouse model of uveoretinitis [Sakai *et al.*, 2006]. To improve nanosteroid biodistribution and half-life, betamethasone was encapsulated by a blend of poly(lactic acid) (PLA) and polyethylene glycol (PEG), which serves to reduce opsonization and sequestration within reticuloendothelial systems [Ishihara *et al.*, 2009b; Ishihara *et al.*, 2009a]. Nanosteroid PEGylation resulted in decreased particle accumulation in the liver and increased accumulation in the spleen 24 hours following intravenous administration. Importantly, the authors observed increased circulating blood levels and increased delivery of PEGylated nanosteroid to the site of inflammation over the same period of time. Matsuo *et al.* [2009] applied the use of PEGylated nanosteroid particles as a treatment strategy in a murine model of asthma. Following intravenous administration, the authors were able to detect nanosteroid delivery to lung tissue as well as enhanced and prolonged betamethasone secretion within the lungs of asthmatic mice compared to untreated or conventional betamethasone-treated mice. Furthermore, a single dose of nanosteroid reduced the overall number of inflammatory cells (e.g., macrophages, eosinophils, neutrophils, and lymphocytes) and reduced IL-13 and IL-4 within bronchoalveolar lavage fluid. Overall, nanosteroid treatment resulted in decreased airway hyperresponsiveness.

5.3.2 Nanocarrier Vaccines as Immune Modulators to Promote T_H1 Responses

In recent years, the incidence of allergic disease has reached epidemic proportions [Holgate, 1999; Umetsu *et al.*, 2002]. Many believe the increased incidence is due to what is known as the hygiene hypothesis, which states that the increased prevalence of allergic diseases can be attributed to decreased exposure to immune-stimulating infections in early childhood [Strachan, 1989]. Decreased exposure to respiratory infections is thought to create an imbalance in T_H1 and T_H2 immunity, allowing a prevalence of T_H2-type responses and, therefore, an increased propensity for asthma. This phenomenon has caused some to speculate whether promoting T_H1-type responses can prevent or alleviate asthma. Distinct cytokines released by T_H1 and T_H2 cells inhibit the generation and activation of each other [Abbas *et al.*, 1996]. For example, T_H1 cells can suppress T_H2 cells via the secretion of IFN-γ whereas T_H2 cells can suppress T_H1 cells through the release of IL-4 and IL-10 (Fig. 5.3). Therefore, promoting T_H1-type responses may inhibit T_H2-type responses and lead to symptomatic relief.

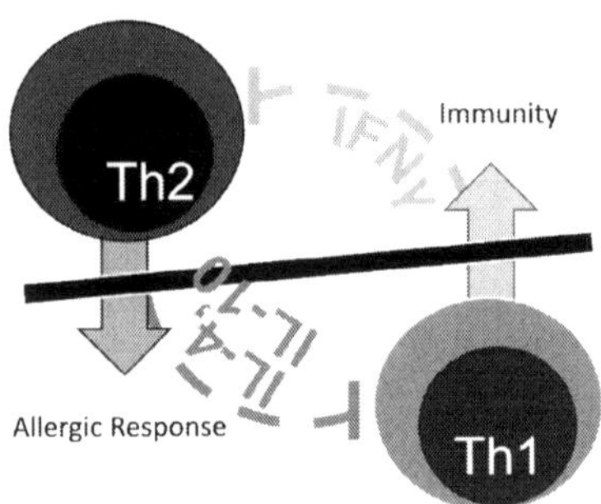

Figure 5.3 The basis for T_H1 skewing as a treatment for asthma. Asthma, as described by the hygiene hypothesis, is associated with a predominately T_H2-type response. T_H2 cells secrete cytokines like IL-4 and IL-10, which inhibit the generation of T_H1 cells. Immunotherapeutic strategies for asthma attempt to alter the T_H2-type microenvironment via the generation of T_H1 cells, which can inhibit T_H2 cell function by the secretion of cytokines such as IFN-γ. See also Color Insert.

One method of inducing T_H1-type responses toward a particular antigen is vaccination. Vaccination entails the delivery of antigen to antigen-presenting cells like DCs. The context in which the antigen is delivered to the antigen-presenting cell determines the type of T cell response elicited toward the antigen of interest. To induce T_H1 skewing, the antigen must be delivered to DCs in the context of certain immune stimulatory signals mediated by what are called vaccine adjuvants. Adjuvants are often made of pathogen-associated molecular patterns (PAMPs), which are molecules associated with different groups of pathogens, such as bacteria and viruses. PAMPs are recognized by cells of the innate immune system via toll-like receptors (TLRs). DCs express an abundance of TLRs, which upon recognition of PAMPs, induce maturation. Maturation of DCs results in the upregulation of costimulatory molecules and expression of cytokines necessary for priming T_H1-type responses.

One such adjuvant that has shown promise is CpG oligodeoxynucleotides (CpG ODN). Unmethylated CpG dinucleotides are an important immune stimulatory component of bacterial DNA and are not present in mammalian DNA. Unmethylated CpG ODNs are recognized by the TLR-9 expressed by professional antigen-presenting cells (e.g., DCs, B cells, and macrophages) and promote T_H1-type and regulatory-type immune responses. The administration of CpG ODNs in combination with allergen has been shown to be effective in reversing established T_H2-mediated eosinophilic airway inflammation. CpG studies have demonstrated reduced allergen-specific IgE antibodies levels, increased T_H1-type cytokines (e.g., IFN-γ and IP-10), and decreased T_H2-type cytokines (e.g., IL-5 and eotaxin) [Kline *et al.*, 2002; Santeliz *et al.*, 2002; Serebrisky *et al.*, 2000]. The immunologic perturbations of CpG ODN treatment have resulted in fewer features of airway remodeling (e.g., subepithelial collagen deposition and goblet cell hyperplasia) and amelioration of airway hyperresponsiveness [Jain *et al.*, 2002]. Interestingly, CpG treatment was found to increase the expression of the regulatory cytokines TGF-β and IL-10 [Jain *et al.*, 2002; Kitagaki *et al.*, 2002], suggesting the activation of regulatory T cells, which can inhibit T_H2-type responses *in vivo* [Cottrez *et al.*, 2000].

Encapsulation of antigen into nanoparticulate carrier systems has been proposed for nasal immunization as a noninvasive means by which to induce immune responses to infectious agents. Chitosan-based particles have been thoroughly studied as carriers for therapeutic molecules and antigens [Amidi *et al.*, 2010] and have been shown to induce humoral responses to a variety of antigens [Baudner *et al.*, 2003; Khatri *et al.*, 2008; Nagamoto *et al.*, 2004]. However, the major limitation of this approach for treating asthma is its propensity to generate strong T_H2-type humoral responses. It is, therefore, ineffective for T_H1 skewing and could potentially even exacerbate symptoms of asthma. To address this drawback, Slütter *et al.* proposed the use of unmethylated CpG ODN as a physical nanoparticle cross-linker in place of tripolyphosphate (TPP) for vaccination against ovalbumin using a chitosan-based nanoparticle platform [Slutter *et al.*, 2010]. They found that incorporation of a CpG cross-linker skewed the immune response against ovalbumin toward a T_H1-type response, with T cells exhibiting high levels of IFN-γ expression following re-exposure to ovalbumin *in vitro*, whereas mice vaccinated with particles using a TPP cross-linker exhibited negligible IFN-γ expression. This study highlights the potential of nanoparticles to alter the T cell microenvironment via the targeting of other immune cells.

5.4 Potential T Cell–Targeted Strategies for Nanoparticle-Based Therapies

Intravenously administered nanoparticles circulate throughout the body and preferentially diffuse into inflamed or neoplastic tissue via the enhanced permeability and retention (EPR) effect [Maeda *et al.*, 2000]. The EPR effect is the property whereby molecules of a certain size have a tendency to accumulate within the vasculature of tumors or inflamed sites to a greater extent than normal tissue. This phenomenon has been well characterized in the case of tumors. Once tumors reach a certain size, they become dependent on the vascular delivery of oxygen and nutrients for survival and continued growth. To stimulate neovascularization, tumor cells secrete growth factors like vascular endothelial growth factor (VEGF). However, unlike physiologic angiogenesis, cancer angiogenesis results in the formation of vasculature with abnormal architecture. Tumor vasculature is characterized by poorly aligned endothelial cells with wide fenestrations, often lacking a smooth muscle layer and adequate lymphatic drainage. These characteristics result in abnormal fluid dynamics within the tumor tissue, allowing nanoparticles to extravasate from the vasculature into tumor tissue through leaky blood vessels, where they are retained for a period of time due to inadequate drainage. Similarly, the vasculature of inflamed tissue is known to be more permeable due to the release of inflammatory cytokines, which allows for the infiltration of immune cells [Laverman *et al.*, 2001]. However, using this nonspecific method of delivery, the percentage of injected dose reaching the target site is usually very low [von Maltzahn *et al.*, 2009] and delivery to nontarget sites, such as the liver and spleen, is comparatively high [Niidome *et al.*, 2006]. To increase delivery specificity, a variety of nanoparticle surface modifications can be made, including conjugation with antibodies, single chain variable fragments (scFvs), tetramers, aptamers, peptides, cytokines, and hormone analogs (Table 5.1).

Table 5.1 Size range of targeting ligands for nanoparticle delivery to T cells

Targeting ligand	Size
Antibody	~155 kDa
Tetramer	~220 kDa
Hormone analog (e.g., Transferrin)	~80 kDa
Single-chain fragment variables	~25 kDa
Aptamer	8–15 kDa
Peptide	~1 kDa

T cells express a variety of unique surface molecules (Fig. 5.4). Targeting these unique molecules offers the potential to increase treatment specificity and perhaps lead to enhanced treatment efficacy. Concurrently, targeted delivery may minimize toxic side effects by decreasing off-target delivery or allowing for the administration of lower drug doses. Current therapies for the treatment of T cell-mediated autoimmune diseases target surface molecules on T cells to directly inhibit their function. These therapeutic strategies include the inhibition of antigen-specific T cells by preventing the formation of the major histocompatibility complex (MHC)–antigen–T cell receptor (TCR) complex, the nonspecific ablation of whole T cell populations using anti-CD4 monoclonal antibodies, and costimulatory molecule inhibition. In addition, T cells also express a variety of unique cytokine and chemokine receptors, as well as adhesion molecules like integrins that may serve as targets.

Depending on the particular target, one can attain different levels of specificity toward a subset of T cells. For instance, targeting a pan T-cell marker like CD3 may not offer much improvement over generalized steroid-based approaches. Asthma is known to be instigated by helper T cells; therefore, treatment specificity may be enhanced by targeting the CD4 molecule. However, while CD4-targeting may increase specificity by targeting a single subset of T cells, it is still relatively nonspecific. Still, treatment specificity may be potentially increased by targeting narrower and narrower populations of T cells with the ultimate goal of targeting a particular allergen-specific T cell. In the following sections, we will explore the potential targeting ligands that may be used to direct nanoparticles to these unique T cell populations (Table 5.2 on next page).

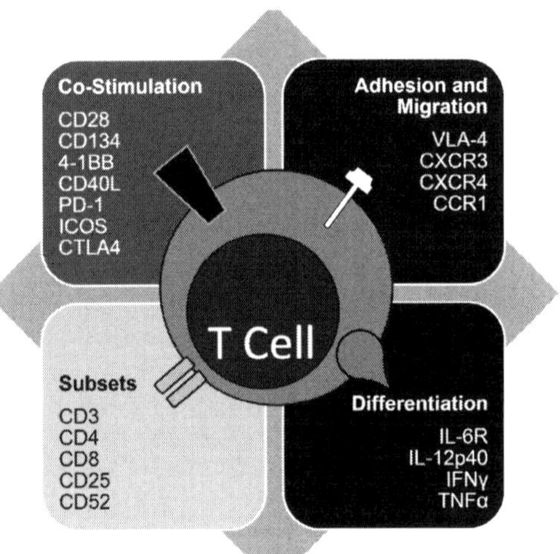

Figure 5.4 Targetable T cell surface molecules. T cells express a variety of unique surface molecules that may serve as targets to increase treatment specificity. Therapeutic strategies include targeting of whole T cell subsets, interference with the TCR complex, blockade of T cell costimulation, obstruction of T cell adhesion and migration, and inhibition of T cell differentiation. See also Color Insert.

Table 5.2 Summary of T cell-targeted strategies using nanoparticles

T cell target	Targeting ligand	Therapeutic molecule	Particle type	Effect	Ref
—	—	Betamethasone	PLGA	Sustained steroid release *in vivo* and decreased inflammation in rat mouse models of arthritis	Ishihara *et al.*, 2005 Higaki *et al.*, 2005
—	—	Betamethasone	PLA/PEG	Sustained steroid release, decreased inflammatory cell infiltration, reduced IL-4 and IL-13 secretion, and decreased airway hyperresponsiveness in the mouse model of asthma	Matsuo *et al.*, 2009
—	—	Ovalbumin and CpG ODN	Chitosan	T_H1-skewed T cell response	Slutter *et al.*, 2010
CD3	Antibody	Doxorubicin	PAMAM	Inhibition of T cell proliferation *in vitro*	Fahmy *et al.*, 2007
CD3	scFv	Plasmid DNA encoding diacylglycerol kinase	PEG/PEI Iron Oxide	Inhibition of T cell proliferation and IL-2 secretion	Chen *et al.*, 2009
TCR	Tetramer	Doxorubicin	PAMAM	Inhibition of antigen-specific T cell proliferation *in vitro* and *in vivo*	Fahmy *et al.*, 2007
4-1BB	Aptamer	—	—	Enhanced T cell proliferation, IFN-γ secretion, and tumor clearance	McNamar *et al.*, 2008
CTLA-4	Aptamer	—	—	Enhanced T cell proliferation and antitumor activity	Santulli *et al.*, 2003
LFA-1	Peptide	—	PLGA	Inhibited T cell LFA-1 interaction with ICAM-1 of lung epithelial cells	Huang *et al.*, 2007

5.5 T Cell-Targeting Ligands

5.5.1 Large Targeting Ligands

5.5.1.1 Antibody conjugates to target T cell surface molecules

Monoclonal antibody therapy is a pharmacologic tool first introduced for the treatment of cancer. Both trastuzamab, which targets human epidermal growth factor receptor 2 (HER2), and rituximab, which targets CD20, have had great success in the treatment of $HER2^+$ breast cancers and hematologic neoplasms such as leukemias and lymphomas, respectively [Cobleigh *et al.*, 1999; Sehn *et al.*, 2005; Vogel *et al.*, 2002]. Building on the success of these therapies, other applications for antibody therapy are being actively pursued, such as the treatment of autoimmune diseases such as rheumatoid arthritis. With an understanding of the basic biology of T cell involvement in autoimmune disease, it is possible to target all aspects of the immune response, from T cell activation within lymph nodes during immunologic synapse formation to T cell differentiation, migration, and cytokine secretion.

Antibody technology has been utilized in the nanomedicine field in an attempt to direct nanoparticles toward certain tumor cell types. For example, monoclonal antibodies specific for EGFR and HER2 have been conjugated to the surface of gold nanoparticles to target $EGFR^+$ and $HER2^+$ breast cancer cells for photothermal therapy [Carpin *et al.*, 2010; Loo *et al.*, 2005; Melancon *et al.*, 2008]. Likewise, antibodies targeting the transferrin receptor have been conjugated to temozolomide-bearing β-poly(L-malic acid) (PMLA) particles for the treatment of glioma [Patil *et al.*, 2010]. Although these studies showed increased nanoparticle tumor specificity *in vitro, in vivo* studies demonstrated only marginal increased tumor localization following antibody conjugation, which was not statistically significant [Melancon *et al.*, 2008].

Antibody conjugation has also been used to direct nanoparticles to immunologic cell types. Antigen-bearing PLGA particles have been targeted to DCs using a monoclonal antibody to DC-SIGN [Cruz *et al.*, 2010]. B cells have been targeted by doxorubicin-loaded liposomes following conjugation with an antibody to CD22 [Loomis *et al.*, 2010]. Likewise, Fahmy *et al* [2007] directed immunosuppressive drugs to T cells *in vitro* using polyamidoamine (PAMAM) particles conjugated with antibody to CD3, a pan T cell marker. In this study, the authors demonstrated that CD3-targeted particles can bind T cells with high specificity. Furthermore, they demonstrated that anti-CD3 conjugated particles bearing doxorubicin could inhibit the proliferation of T cells *in vitro*. However, the major limitation of this approach is that CD3 is expressed by all T cells and targeting this molecule would negate the effect of using antibody-conjugation strategies to increase specificity and target only disease-provoking T cell subsets.

There are many unique surface molecules expressed by T cells that may be exploited for antibody-based strategies (Table 5.3). Entire T cell populations may be targeted using antibodies to CD4 or CD8 to direct particles to helper or cytotoxic T cell populations, respectively. Upon activation, T cells upregulate the expression of CD25, the receptor for the T cell growth cytokine IL-2. Therefore, targeting CD25 may be a way to direct particles to activated T cells. CD25 is also expressed at high levels by regulatory T cells. Hence, drugs that stimulate T_{reg} function may be delivered via CD25 targeting to inhibit immune responses. Like CD25, the expression of costimulatory molecules necessary for the complete activation of T cells is also upregulated upon initial activation. Therefore, targeting molecules such as CD134, 4-1BB, or CD40L represents another strategy for nanoparticle delivery to activated T cells.

Antibodies may be used in their native form or as antibody fragments. Antibody fragments include antigen–binding fragments (Fab'), dimers of antigen-binding fragments ($F(ab')_2$), and single-chain fragment variables (scFv) [Peer *et al.*, 2007]. Specifically, scFvs are fusion proteins joining variable heavy (V_H) and light (V_L) chains of an antibody with a short linker peptide. This strategy allows one to maintain the specificity of an antibody while reducing the size of the targeting ligand by eliminating conserved antibody structural components. This technology has been used extensively to generate T cell chimeric antigen receptors that allow T cell recognition of tumor-associated antigens in an HLA-independent manner [Leen *et al.*, 2007]. Like antibodies, scFvs may be conjugated to nanoparticle surfaces in an attempt to target specific cells or tissues *in vivo*. In recent studies, liposomal nanoparticles have been conjugated with scFvs for the delivery of siRNA and miRNA to B16F10 melanoma tumor cells *in vivo* [Chen *et al.*, 2010]. Additionally, an scFv specific for CD3 has been used to deliver immunosuppressive genes, like diacylglycerol kinases, to T cell *in vitro* [Chen *et al.*, 2009]. T cells are inherently difficult to transfect. Interestingly, this study demonstrated that conjugation of PEGylated polyethylenimine-coated iron oxide nanoparticles with a scFv to CD3 allowed for the efficient transfection of a rat T cell line with plasmid DNA. By incorporating plasmid DNA encoding diacylglycerol kinase, a protein that impairs TCR signaling, the authors were able to inhibit the proliferation and IL-2 secretion of stimulated T cells.

Table 5.3 Current and potential antibody-based T cell targeting strategies

Mechanism of action	Surface target	Cell population	Antibody
Blockade and depletion of T cell populations	CD3	Pan-T cell	Visilizumab
	CD4	T_{helper} and T_{reg}	
	CD8	$T_{Cytotoxic}$	
	CD25 (IL-2R)	Activated T cells and T_{reg}	Daclizumab, Basiliximab
	CD52	T cells, B cells, NK cells, dendritic cells, monocytes, and granulocytes	Alemtuzumab
Blockade of the immunological synapse	CD11 (LFA1)	——	Efalizumab
Prevention of costimulation	CD28 CD134 (OX40) 4-1BB CD40L PD-1 ICOS	——	
	CTLA4		Abatacept
Blockade of T cell adhesion and migration	VLA-4	All leukocytes except neutrophils	Natalizumab
	CD11 (LFA-1)		Efalizumab
Prevention of T cell differentiation	IL-6R	T_H17	Tocilizumab
	IL-12p40	T_H1 and T_H17	Ustekinumab
	IFN-γ	T_H1	Fontolizumab
Blockade of proinflammatory cytokines	TNFα	——	Infliximab

5.5.1.2 TCR-targeted strategies

The specificity of the T cell response is mediated by TCR recognition of antigen presented in the context of MHC molecules. Thus, antigen-specific T cell populations may be targeted by peptide-MHC complexes. Such a strategy has been used extensively for the detection of antigen-specific T cells in laboratory settings and has been proposed as a potential means of inducing T cell anergy in patients [Altman *et al.*, 1996; Howard *et al.*, 1999]. Soluble monomeric peptide-MHC complexes have an inherently low TCR affinity [Corr *et al.*, 1994; Sykulev *et al.*, 1994] and must be constructed into dimeric and tetrameric forms to enhance TCR-binding avidity. In an attempt to utilize this approach to target a particular antigen-reactive T cell population, researchers have generated peptide-MHC chimeras with immunosuppressive drugs, such as doxorubicin [Casares *et al.*, 2001].

However, chemical conjugation of therapeutic molecules to antigens that can be recognized and bound by the TCR may be difficult and result in hindered antigen presentation. To address this problem, researchers have functionalized doxorubicin encapsulated PAMAM nanoparticles with MHC molecules and observed selective TCR binding to and subsequent internalization of particles [Fahmy *et al.*, 2007]. Furthermore, these particles were able to inhibit the expansion of antigen-specific T cells *in vitro* and even more impressively *in vivo*.

Although TCR targeting has the potential to direct drugs toward a clonal population of T cells, the use of tetramers has certain limitations. In complex diseases such as asthma, the instigating antigen is often unknown. Furthermore, in the event that a particular antigen is identified, it is often difficult to determine the particular epitopes recognized by the TCR. Epitope identification is often a cumbersome process and varies significantly according to a patient's particular human leukocyte antigen haplotype. The various limitations of antigen and epitope identification thus prevent the widespread clinical application of tetramer-based therapies.

Thus far, we have discussed nanoparticle delivery to T cells by targeting unique surface structures using tetramers and antibodies. Although these targeting ligands have an established track record of proven specificity, the use of these macromolecular structures may interfere with nanoparticle bioavailability and cellular uptake, eliminating the benefits of nano-based approaches. It has been proposed that smaller molecules may enhance specificity while maintaining the benefits of using a nanoparticle for delivery. We will now focus on the use of smaller targeting ligands that may be more appropriate for nanoparticle drug delivery.

5.5.2 Small Targeting Ligands

5.5.2.1 Aptamers

Aptamers are single-stranded nucleic acid ligands that represent a new class of targeting agents. They are globular molecules that do not encode any genetic information or interfere with mRNA translation and function through high affinity binding to target proteins or small molecules. Aptamers can be generated that bind essentially any protein through repetitive *in vitro* selection techniques using short RNA libraries [Gold, 1995; White *et al.*, 2000] and can bind with equivalent or greater specificity and avidity than monoclonal antibodies.

Aptamer-based technologies offer a variety of benefits over monoclonal antibody approaches [Hicke *et al.*, 2000]. Aptamers are chemically synthesized, which allows for a more expedited regulatory approval process compared to the cell-based synthesis process of monoclonal antibodies. Furthermore, aptamers are predicted to have less potential to induce antiaptamer immune responses, which are a common problem seen with the use of monoclonal antibodies derived from different species. Such a property allows for repeat dosing in the clinical setting. Little or no immunogenicity was observed in animal studies analyzing the pharmacokinetics and toxicity of an anti-VEGF aptamer [Drolet *et al.*, 2000], and several clinical trials using aptamer-targeted strategies have been conducted [Dyke *et al.*, 2006; Mongelard *et al.*, 2010]. Finally, antibodies are thought to be inherently handicapped by their large size (~155 kDa), which results in inefficient tissue penetration. In comparison, aptamers (~8–15 kDa) are roughly 10% the size of antibodies. This smaller size allows for rapid tissue penetration and is perhaps more suitable for conjugation to nanoscale particles.

Several studies have demonstrated the ability of aptamers to bind surface molecules on T cells both *in vitro* and *in vivo* to enhance their function. McNamara *et al.* [2008] developed an aptamer that could recognize and bind to 4-1BB, a costimulatory molecule expressed by activated T cells. The authors further demonstrated that a 4-1BB-specific aptamer could enhance the proliferation of and IFN-γ secretion by stimulated T cells. Importantly, intratumoral administration of the aptamer resulted in tumor rejection and increased survival in mice, presumably due to enhanced activation of antitumor T cells. In a similar fashion, Santulli-Marotto *et al.* [2003] developed an aptamer specific for CTLA-4, an inhibitory molecule expressed by activated T cells. CTLA-4 recognizes and binds to the costimulatory molecules CD80 and CD86 with higher affinity than CD28. Unlike CD28, which functions as a T cell costimulatory molecule, CTLA-4 delivers a negative signal to attenuate T cell responses [Alegre *et al.*, 2001]. The authors developed an aptamer that could bind to and inhibit CTLA-4 signaling, thereby resulting in enhanced T cell proliferation *in vitro* and augmented antitumor responses *in vivo*. Additionally, the administration of the inhibitory CTLA-4 aptamer enhanced the immune response to tumor antigens following DC vaccination. Interestingly, both of these studies demonstrate the ability of aptamers to have agonistic and antagonistic properties. Therefore, in addition to functioning as targeting ligands to direct nanoparticles to T cells *in vivo*, they may also have the added benefit of further enhancing or inhibiting immune responses.

5.5.2.2 Peptides

Small peptides that can bind specific cell-surface receptors with high affinity and selectivity have been used to target drugs and imaging agents to certain tissues such as tumors and have been studied as a means of targeted nanoparticle delivery [Juliano *et al.*, 2009]. Conjugation of nanoparticles with natural polypeptide ligands, like transferrin, allows for specific nanoparticle delivery to cells expressing the particular cell-surface receptor and can promote cellular uptake of particles *in vivo* by receptor-mediated endocytosis [Heidel *et al.*, 2007; Sahoo *et al.*, 2004; Widera *et al.*, 2003].

Natural polypeptide ligands are large molecules (i.e., transferrin is ~80 kDa [Roberts *et al.*, 1966]) and may not be ideal for nanoscale therapeutics. Using phage display, RNA display, and bacterial or yeast display technologies, novel short peptide sequences can be identified from large peptide libraries that bind a particular cell-surface receptor [Huang *et al.*, 2007; Jostock *et al.*, 2005; Kehoe *et al.*, 2005]. These screening techniques usually reveal short peptides consisting of 6–12 amino acids. One particular class of cell-targeting peptide that binds integrins contains an Arg-Gly-Asp tripeptide sequence known as an Arg-Gly-Asp (RGD) motif. Unmodified RGD motifs bind integrins with low affinity. To increase binding affinity, the RGD moiety can be modified or made into a cyclic form. Furthermore, via the use of short cationic or hydrophobic peptides, certain sequences can lead to peptide internalization after binding. Conjugation of nanoparticles with cell-penetrating peptide sequences, such as transactivator of transcription (TAT) fragments derived from HIV TAT, may thereby allow intracellular delivery of drug-bearing nanoparticles.

Integrins may also present a therapeutic target. Integrins play in important role in leukocyte adhesion to endothelium and transmigration from the vasculature into inflamed tissues. Leukocytes express the β_2 integrin leukocyte function associated antigen-1 (LFA-1). LFA-1 binds to intracellular adhesion molecule-1 (ICAM-1) expressed on endothelial cells. ICAM-1 is upregulated by endothelium in the presence of inflammatory cytokines such as IL-1, IL-4, IL-6, TNF-α, and IFN-γ. Therefore, targeting integrins important for leukocyte trafficking, like LFA-1, represents a way to deliver therapeutic particles to T cells as they are entering into inflamed tissue. One study utilized a small peptide antagonist to LFA-1, called cIBR, to bind nanoparticles to the surface of T cells [Chittasupho *et al.*, 2010]. The authors demonstrated that conjugating the nanoparticle surface with cIBR leads to nanoparticle binding to the T cell surface and subsequent internalization. Furthermore, cIBR binding to the T cell surface inhibited LFA-1 interaction with ICAM-1 expressed by lung epithelial cells *in vitro* by either direct antagonistic effects or by decreasing the availability of LFA-1 on the T cell surface due to receptor-mediated endocytosis. Such a strategy represents a dual therapeutic approach by both preventing T cell migration to sites of inflammation and ensuring concurrent delivery of anti-inflammatory drugs.

5.6 Alternative Approaches

5.6.1 Chemokine Receptor-Targeted Strategies

The expression of unique chemokine receptor profiles may be a means by which to selectively target specific T cell populations. Inflammatory disorders are characterized by excessive infiltration of leukocytes to sites of inflammation. Leukocyte recruitment is controlled by several factors, including proinflammatory cytokines, adhesion molecules, and chemotactic cytokines known as chemokines. Chemokines signal through G-protein-coupled receptors on leukocytes to control homeostatic trafficking and homing to sites infection. Certain chemokine receptors are constitutively expressed by T cells and are involved in basal T cell trafficking, while the expression of others is inducibly expressed by certain inflammatory processes [Proudfoot, 2002]. T cell subsets express unique chemokine receptor profiles [Sallusto *et al.*, 1998; Sallusto *et al.*, 1999; Zingoni *et al.*, 1998]. In regard to inflammatory airway diseases such as asthma, the expression of various chemokine receptors by T cells has been identified (Table 5.4).

Table 5.4 Chemokine receptor expression by T cells implicated in asthma pathogenesis [Bochner *et al.*, 2003; Gonzalo *et al.*, 2000; Lukacs *et al.*, 2002; Panina-Bordignon *et al.*, 2001]

T cell subset	Chemokine receptor	Ligand
T_H1	CCR5	CCL4 (MIP-1β)
T_H2	CCR4	CCL17 (TARC)
	CCR8	CCL1 (I-309)
	CXCR4	CXCL12 (SDF-1) CCL11
	CCR3	(Eotaxin)

Chemokine biology is a relatively new field, and there are a limited number of identified neutralizing antibodies and small molecule inhibitors with *in vivo* activity. Antibodies to both CXCR3 and CXCR4 as well as a novel peptide to CCR1 have been identified [Hancock *et al.*, 2000; Liang *et al.*, 2000]. Furthermore, a variety of small molecule inhibitors have been found to bind chemokine receptors, such as the CCR5 inhibitor TAK-779 [Baba *et al.*, 1999]. However, to date, there are no studies demonstrating cytokine receptor-targeted nanoparticle delivery.

5.7 Summary

This chapter has attempted to demonstrate how nanotechnology may be applied to diseases where T cells play an important or initiating role. We have chosen to largely focus on allergic asthma for the simple reason that T cell involvement in the pathogenesis of this disease is well defined. However, similar approaches may be applied toward other T cell–mediated diseases with widely varying etiologies such as chronic obstructive pulmonary disease, rheumatoid arthritis, and multiple sclerosis. The application of nanotechnologies to disease is an interdisciplinary venture requiring knowledge of both the disease mechanism and the nanoparticles. With a detailed understanding of disease pathophysiology, nanomedicine may be tailored to direct therapeutic molecules to disease-instigating cell populations. In the case of asthma, nanotechnology can be used to localize treatments, decreasing the risk of severe side effects. The addition of targeting moieties to nanoparticles can further optimize and enhance treatment specificity for a particular population of cells. In this chapter, we have discussed the use of nanoparticles to deliver and promote the sustained release of steroids at sites of inflammation. We have also evaluated the potential of nanoparticles to alter the disease microenvironment by skewing T_H2-type responses toward the T_H1-type through a nanoparticle-based vaccine. Finally, we explored the various nanoparticle modifications that can potentially be developed to allow specific targeting of various T cell subsets. The use of nanotechnology in future treatments of asthma may enhance the efficacy and specificity of asthma treatment, decrease the incidence of severe side effects secondary to asthma treatment, and better target the pathogenic mediators of atopic asthma.

References

1. Abbas, A. K., Murphy, K. M., and Sher, A. (1996). Functional diversity of helper T lymphocytes, *Nature*, 383, pp. 787–793.
2. Akdis, M., Verhagen, J., Taylor, A., Karamloo, F., Karagiannidis, C., Crameri, R., Thunberg, S., Deniz, G., Valenta, R., Fiebig, H., Kegel, C., Disch, R., Schmidt-Weber, C. B., Blaser, K., and Akdis, C. A. (2004). Immune responses in healthy and allergic individuals are characterized by a fine balance between allergen-specific T regulatory 1 and T helper 2 cells, *J. Exp. Med.*, 199, pp. 1567–1575.
3. Alegre, M. L., Frauwirth, K. A., and Thompson, C. B. (2001). T-cell regulation by CD28 and CTLA-4, *Nat. Rev. Immunol.*, 1, pp. 220–228.
4. Alexander, A. G., Barnes, N. C., and Kay, A. B. (1992). Trial of cyclosporin in corticosteroid-dependent chronic severe asthma, *Lancet*, 339, pp. 324–328.

5. Altman, J. D., Moss, P. A., Goulder, P. J., Barouch, D. H., Heyzer-Williams, M. G., Bell, J. I., McMichael, A. J., and Davis, M. M. (1996). Phenotypic analysis of antigen-specific T lymphocytes, *Science*, 274, pp. 94–96.
6. Amidi, M., Mastrobattista, E., Jiskoot, W., and Hennink, W. E. (2010). Chitosan-based delivery systems for protein therapeutics and antigens, *Adv. Drug Deliv. Rev.*, 62, pp. 59–82.
7. Baba, M., Nishimura, O., Kanzaki, N., Okamoto, M., Sawada, H., Iizawa, Y., Shiraishi, M., Aramaki, Y., Okonogi, K., Ogawa, Y., Meguro, K., and Fujino, M. (1999). A small-molecule, nonpeptide CCR5 antagonist with highly potent and selective anti-HIV-1 activity, *Proc. Natl. Acad. Sci. U.S.A.*, 96, pp. 5698–5703.
8. Bacharier, L. B., Jabara, H., and Geha, R. S. (1998). Molecular mechanisms of immunoglobulin E regulation, *Int. Arch. Allergy Immunol.*, 115, pp. 257–269.
9. Barnes, P. J. (1995). Inhaled glucocorticoids for asthma, *N. Engl. J. Med.*, 332, pp. 868–875.
10. Barnes, P. J., and Adcock, I. M. (1998). Transcription factors and asthma, *Eur. Respir. J.*, 12, pp. 221–234.
11. Barnes, P. J., and Adcock, I. M. (2003). How do corticosteroids work in asthma?, *Ann. Intern. Med.*, 139, pp. 359–370.
12. Baudner, B. C., Giuliani, M. M., Verhoef, J. C., Rappuoli, R., Junginger, H. E., and Giudice, G. D. (2003). The concomitant use of the LTK63 mucosal adjuvant and of chitosan-based delivery system enhances the immunogenicity and efficacy of intranasally administered vaccines, *Vaccine*, 21, pp. 3837–3844.
13. Bochner, B. S., Hudson, S. A., Xiao, H. Q., and Liu, M. C. (2003). Release of both CCR4-active and CXCR3-active chemokines during human allergic pulmonary late-phase reactions, *J. Allergy Clin. Immunol.*, 112, pp. 930–934.
14. Brusselle, G. G., Kips, J. C., Tavernier, J. H., van der Heyden, J. G., Cuvelier, C. A., Pauwels, R. A., and Bluethmann, H. (1994). Attenuation of allergic airway inflammation in IL-4 deficient mice, *Clin. Exp. Allergy*, 24, pp. 73–80.
15. Burrows, B., Martinez, F. D., Halonen, M., Barbee, R. A., and Cline, M. G. (1989). Association of asthma with serum IgE levels and skin-test reactivity to allergens, *N. Engl. J. Med.*, 320, pp. 271–277.
16. Busse, W. W., and Lemanske, R. F., Jr. (2001). Asthma, *N. Engl. J. Med.*, 344, pp. 350–362.
17. Carpin, L. B., Bickford, L. R., Agollah, G., Yu, T. K., Schiff, R., Li, Y., and Drezek, R. A. (2010). Immunoconjugated gold nanoshell-mediated

photothermal ablation of trastuzumab-resistant breast cancer cells, *Breast Cancer Res. Treat.*, 125(1), pp. 27–34.

18. Casares, S., Stan, A. C., Bona, C. A., and Brumeanu, T. D. (2001). Antigen-specific downregulation of T cells by doxorubicin delivered through a recombinant MHC II--peptide chimera, *Nat. Biotechnol.*, 19, pp. 142–147.
19. Chen, G., Chen, W., Wu, Z., Yuan, R., Li, H., Gao, J., and Shuai, X. (2009). MRI-visible polymeric vector bearing CD3 single chain antibody for gene delivery to T cells for immunosuppression, *Biomaterials*, 30, pp. 1962–1970.
20. Chen, Y., Zhu, X., Zhang, X., Liu, B., and Huang, L. (2010). Nanoparticles modified with tumor-targeting scFv deliver siRNA and miRNA for cancer therapy, *Mol. Ther.*, 18, pp. 1650–1656.
21. Chittasupho, C., Manikwar, P., Krise, J. P., Siahaan, T. J., and Berkland, C. (2010). cIBR effectively targets nanoparticles to LFA-1 on acute lymphoblastic T cells, *Mol. Pharm.*, 7, pp. 146–155.
22. Cobleigh, M. A., Vogel, C. L., Tripathy, D., Robert, N. J., Scholl, S., Fehrenbacher, L., Wolter, J. M., Paton, V., Shak, S., Lieberman, G., and Slamon, D. J. (1999). Multinational study of the efficacy and safety of humanized anti-HER2 monoclonal antibody in women who have HER2-overexpressing metastatic breast cancer that has progressed after chemotherapy for metastatic disease, *J. Clin. Oncol.*, 17, pp. 2639–2648.
23. Cohn, L., Tepper, J. S., and Bottomly, K. (1998). IL-4-independent induction of airway hyperresponsiveness by Th2, but not Th1, cells, *J. Immunol.*, 161, pp. 3813–3816.
24. Corr, M., Slanetz, A. E., Boyd, L. F., Jelonek, M. T., Khilko, S., al-Ramadi, B. K., Kim, Y. S., Maher, S. E., Bothwell, A. L., and Margulies, D. H. (1994). T cell receptor-MHC class I peptide interactions: affinity, kinetics, and specificity, *Science*, 265, pp. 946–949.
25. Corry, D. B., Folkesson, H. G., Warnock, M. L., Erle, D. J., Matthay, M. A., Wiener-Kronish, J. P., and Locksley, R. M. (1996). Interleukin 4, but not interleukin 5 or eosinophils, is required in a murine model of acute airway hyperreactivity, *J. Exp. Med.*, 183, pp. 109–117.
26. Cottrez, F., Hurst, S. D., Coffman, R. L., and Groux, H. (2000). T regulatory cells 1 inhibit a Th2-specific response *in vivo*, *J. Immunol.*, 165, pp. 4848–4853.
27. Coyle, A. J., Le, G. G., Bertrand, C., Tsuyuki, S., Heusser, C. H., Kopf, M., and Anderson, G. P. (1995). Interleukin-4 is required for the induction of lung Th2 mucosal immunity, *Am. J. Respir. Cell Mol. Biol.*, 13, pp. 54–59.

28. Cruz, L. J., Tacken, P. J., Fokkink, R., Joosten, B., Stuart, M. C., Albericio, F., Torensma, R., and Figdor, C. G. (2010). Targeted PLGA nano- but not microparticles specifically deliver antigen to human dendritic cells via DC-SIGN *in vitro*, *J. Controlled Release*, 144, pp. 118–126.

29. Dahl, R. (2006). Systemic side effects of inhaled corticosteroids in patients with asthma, *Respir. Med.*, 100, pp. 1307–1317.

30. Drolet, D. W., Nelson, J., Tucker, C. E., Zack, P. M., Nixon, K., Bolin, R., Judkins, M. B., Farmer, J. A., Wolf, J. L., Gill, S. C., and Bendele, R. A. (2000). Pharmacokinetics and safety of an anti-vascular endothelial growth factor aptamer (NX1838) following injection into the vitreous humor of rhesus monkeys, *Pharm. Res.*, 17, pp. 1503–1510.

31. Dyke, C. K., Steinhubl, S. R., Kleiman, N. S., Cannon, R. O., Aberle, L. G., Lin, M., Myles, S. K., Melloni, C., Harrington, R. A., Alexander, J. H., Becker, R. C., and Rusconi, C. P. (2006). First-in-human experience of an antidote-controlled anticoagulant using RNA aptamer technology: a phase 1a pharmacodynamic evaluation of a drug-antidote pair for the controlled regulation of factor IXa activity, *Circulation*, 114, pp. 2490–2497.

32. Fahmy, T. M., Schneck, J. P., and Saltzman, W. M. (2007). A nanoscopic multivalent antigen-presenting carrier for sensitive detection and drug delivery to T cells, *Nanomedicine*, 3, pp. 75–85.

33. Foster, P. S., Hogan, S. P., Ramsay, A. J., Matthaei, K. I., and Young, I. G. (1996). Interleukin 5 deficiency abolishes eosinophilia, airways hyperreactivity, and lung damage in a mouse asthma model, *J. Exp. Med.*, 183, pp. 195–201.

34. Gavett, S. H., Chen, X., Finkelman, F., and Wills-Karp, M. (1994). Depletion of murine CD4+ T lymphocytes prevents antigen-induced airway hyperreactivity and pulmonary eosinophilia, *Am. J. Respir. Cell Mol. Biol.*, 10, pp. 587–593.

35. Gold, L. (1995). Oligonucleotides as research, diagnostic, and therapeutic agents, *J. Biol. Chem.*, 270, pp. 13581–13584.

36. Gonzalo, J. A., Lloyd, C. M., Peled, A., Delaney, T., Coyle, A. J., and Gutierrez-Ramos, J. C. (2000). Critical involvement of the chemotactic axis CXCR4/stromal cell-derived factor-1 alpha in the inflammatory component of allergic airway disease, *J. Immunol.*, 165, pp. 499–508.

37. Grunig, G., Warnock, M., Wakil, A. E., Venkayya, R., Brombacher, F., Rennick, D. M., Sheppard, D., Mohrs, M., Donaldson, D. D., Locksley, R. M., and Corry, D. B. (1998). Requirement for IL-13 independently of IL-4 in experimental asthma, *Science*, 282, pp. 2261–2263.

38. Hancock, W. W., Lu, B., Gao, W., Csizmadia, V., Faia, K., King, J. A., Smiley, S. T., Ling, M., Gerard, N. P., and Gerard, C. (2000). Requirement of the

chemokine receptor CXCR3 for acute allograft rejection, *J. Exp. Med.*, 192, pp. 1515–1520.

39. Heidel, J. D., Yu, Z., Liu, J. Y., Rele, S. M., Liang, Y., Zeidan, R. K., Kornbrust, D. J., and Davis, M. E. (2007). Administration in non-human primates of escalating intravenous doses of targeted nanoparticles containing ribonucleotide reductase subunit M2 siRNA, *Proc. Natl. Acad. Sci. U.S.A.*, 104, pp. 5715–5721.

40. Herrick, C. A., and Bottomly, K. (2003). To respond or not to respond: T cells in allergic asthma, *Nat. Rev. Immunol.*, 3, pp. 405–412.

41. Hicke, B. J., and Stephens, A. W. (2000). Escort aptamers: a delivery service for diagnosis and therapy, *J. Clin. Invest*, 106, pp. 923–928.

42. Hodge, M. R., Ranger, A. M., Charles de la, B. F., Hoey, T., Grusby, M. J., and Glimcher, L. H. (1996). Hyperproliferation and dysregulation of IL-4 expression in NF-ATp-deficient mice, *Immunity*, 4, pp. 397–405.

43. Hodge, M. R., Rooney, J. W., and Glimcher, L. H. (1995). The proximal promoter of the IL-4 gene is composed of multiple essential regulatory sites that bind at least two distinct factors, *J. Immunol.*, 154, pp. 6397–6405.

44. Holgate, S. T. (1999). The epidemic of allergy and asthma, *Nature*, 402, pp. B2-B4.

45. Holgate, S. T., and Polosa, R. (2008). Treatment strategies for allergy and asthma, *Nat. Rev. Immunol.*, 8, pp. 218–230.

46. Howard, M. C., Spack, E. G., Choudhury, K., Greten, T. F., and Schneck, J. P. (1999). MHC-based diagnostics and therapeutics - clinical applications for disease-linked genes, *Immunol. Today*, 20, pp. 161–165.

47. Huang, B. C., and Liu, R. (2007). Comparison of mRNA-display-based selections using synthetic peptide and natural protein libraries, *Biochemistry*, 46, pp. 10102–10112.

48. Ishihara, T., Kubota, T., Choi, T., and Higaki, M. (2009a). Treatment of experimental arthritis with stealth-type polymeric nanoparticles encapsulating betamethasone phosphate, *J. Pharmacol. Exp. Ther.*, 329, pp. 412–417.

49. Ishihara, T., Takahashi, M., Higaki, M., and Mizushima, Y. (2009b). Efficient encapsulation of a water-soluble corticosteroid in biodegradable nanoparticles, *Int. J. Pharm.*, 365, pp. 200–205.

50. Jain, V. V., Kitagaki, K., Businga, T., Hussain, I., George, C., O'shaughnessy, P., and Kline, J. N. (2002). CpG-oligodeoxynucleotides inhibit airway remodeling in a murine model of chronic asthma, *J. Allergy Clin. Immunol.*, 110, pp. 867–872.

51. Jostock, T., and Dubel, S. (2005). Screening of molecular repertoires by microbial surface display, *Comb. Chem. High Throughput. Screen.*, 8, pp. 127–133.

52. Juliano, R. L., Alam, R., Dixit, V., and Kang, H. M. (2009). Cell-targeting and cell-penetrating peptides for delivery of therapeutic and imaging agents, *Wiley. Interdiscip. Rev. Nanomed. Nanobiotechnol.*, 1, pp. 324–335.

53. Kearley, J., Barker, J. E., Robinson, D. S., and Lloyd, C. M. (2005). Resolution of airway inflammation and hyperreactivity after *in vivo* transfer of CD4+CD25+ regulatory T cells is interleukin 10 dependent, *J. Exp. Med.*, 202, pp. 1539–1547.

54. Kehoe, J. W., and Kay, B. K. (2005). Filamentous phage display in the new millennium, *Chem. Rev.*, 105, pp. 4056–4072.

55. Khatri, K., Goyal, A. K., Gupta, P. N., Mishra, N., and Vyas, S. P. (2008). Plasmid DNA loaded chitosan nanoparticles for nasal mucosal immunization against hepatitis B, *Int. J. Pharm.*, 354, pp. 235–241.

56. Kim, J., Woods, A., Becker-Dunn, E., and Bottomly, K. (1985). Distinct functional phenotypes of cloned Ia-restricted helper T cells, *J. Exp. Med.*, 162, pp. 188–201.

57. Kitagaki, K., Jain, V. V., Businga, T. R., Hussain, I., and Kline, J. N. (2002). Immunomodulatory effects of CpG oligodeoxynucleotides on established th2 responses, *Clin. Diagn. Lab Immunol.*, 9, pp. 1260–1269.

58. Kline, J. N., Kitagaki, K., Businga, T. R., and Jain, V. V. (2002). Treatment of established asthma in a murine model using CpG oligodeoxynucleotides, *Am. J. Physiol Lung Cell Mol. Physiol*, 283, pp. L170-L179.

59. Kung, T. T., Stelts, D. M., Zurcher, J. A., Adams, G. K., III, Egan, R. W., Kreutner, W., Watnick, A. S., Jones, H., and Chapman, R. W. (1995). Involvement of IL-5 in a murine model of allergic pulmonary inflammation: prophylactic and therapeutic effect of an anti-IL-5 antibody, *Am. J. Respir. Cell Mol. Biol.*, 13, pp. 360–365.

60. Langrish, C. L., Chen, Y., Blumenschein, W. M., Mattson, J., Basham, B., Sedgwick, J. D., McClanahan, T., Kastelein, R. A., and Cua, D. J. (2005). IL-23 drives a pathogenic T cell population that induces autoimmune inflammation, *J. Exp. Med.*, 201, pp. 233–240.

61. Laverman, P., Dams, E. T., Storm, G., Hafmans, T. G., Croes, H. J., Oyen, W. J., Corstens, F. H., and Boerman, O. C. (2001). Microscopic localization of PEG-liposomes in a rat model of focal infection, *J. Controlled Release*, 75, pp. 347–355.

62. Lee, H. J., Matsuda, I., Naito, Y., Yokota, T., Arai, N., and Arai, K. (1994). Signals and nuclear factors that regulate the expression of interleukin-

4 and interleukin-5 genes in helper T cells, *J. Allergy Clin. Immunol.*, 94, pp. 594–604.

63. Lee, J. J., McGarry, M. P., Farmer, S. C., Denzler, K. L., Larson, K. A., Carrigan, P. E., Brenneise, I. E., Horton, M. A., Haczku, A., Gelfand, E. W., Leikauf, G. D., and Lee, N. A. (1997). Interleukin-5 expression in the lung epithelium of transgenic mice leads to pulmonary changes pathognomonic of asthma, *J. Exp. Med.*, 185, pp. 2143–2156.

64. Leen, A. M., Rooney, C. M., and Foster, A. E. (2007). Improving T cell therapy for cancer, *Annu. Rev. Immunol.*, 25, pp. 243–265.

65. Liang, M., Mallari, C., Rosser, M., Ng, H. P., May, K., Monahan, S., Bauman, J. G., Islam, I., Ghannam, A., Buckman, B., Shaw, K., Wei, G. P., Xu, W., Zhao, Z., Ho, E., Shen, J., Oanh, H., Subramanyam, B., Vergona, R., Taub, D., Dunning, L., Harvey, S., Snider, R. M., Hesselgesser, J., Morrissey, M. M., and Perez, H. D. (2000). Identification and characterization of a potent, selective, and orally active antagonist of the CC chemokine receptor-1, *J. Biol. Chem.*, 275, pp. 19000–19008.

66. Ling, E. M., Smith, T., Nguyen, X. D., Pridgeon, C., Dallman, M., Arbery, J., Carr, V. A., and Robinson, D. S. (2004). Relation of CD4+CD25+ regulatory T-cell suppression of allergen-driven T-cell activation to atopic status and expression of allergic disease, *Lancet*, 363, pp. 608–615.

67. Loo, C., Lowery, A., Halas, N., West, J., and Drezek, R. (2005). Immunotargeted nanoshells for integrated cancer imaging and therapy, *Nano. Lett.*, 5, pp. 709–711.

68. Loomis, K., Smith, B., Feng, Y., Garg, H., Yavlovich, A., Campbell-Massa, R., Dimitrov, D. S., Blumenthal, R., Xiao, X., and Puri, A. (2010). Specific targeting to B cells by lipid-based nanoparticles conjugated with a novel CD22-ScFv, *Exp. Mol. Pathol.*, 88, pp. 238–249.

69. Lukacs, N. W., Berlin, A., Schols, D., Skerlj, R. T., and Bridger, G. J. (2002). AMD3100, a CxCR4 antagonist, attenuates allergic lung inflammation and airway hyperreactivity, *Am. J. Pathol.*, 160, pp. 1353–1360.

70. Maeda, H., Wu, J., Sawa, T., Matsumura, Y., and Hori, K. (2000). Tumor vascular permeability and the EPR effect in macromolecular therapeutics: a review, *J. Controlled Release*, 65, pp. 271–284.

71. Melancon, M. P., Lu, W., Yang, Z., Zhang, R., Cheng, Z., Elliot, A. M., Stafford, J., Olson, T., Zhang, J. Z., and Li, C. (2008). *In vitro* and *in vivo* targeting of hollow gold nanoshells directed at epidermal growth factor receptor for photothermal ablation therapy, *Mol. Cancer Ther.*, 7, pp. 1730–1739.

72. Molet, S., Hamid, Q., Davoine, F., Nutku, E., Taha, R., Page, N., Olivenstein, R., Elias, J., and Chakir, J. (2001). IL-17 is increased in asthmatic airways

and induces human bronchial fibroblasts to produce cytokines, *J. Allergy Clin. Immunol.*, 108, pp. 430–438.

73. Mongelard, F., and Bouvet, P. (2010). AS-1411, a guanosine-rich oligonucleotide aptamer targeting nucleolin for the potential treatment of cancer, including acute myeloid leukemia, *Curr. Opin. Mol. Ther.*, 12, pp. 107–114.
74. Mosmann, T. R., Cherwinski, H., Bond, M. W., Giedlin, M. A., and Coffman, R. L. (1986). Two types of murine helper T cell clone. I. Definition according to profiles of lymphokine activities and secreted proteins, *J. Immunol.*, 136, pp. 2348–2357.
75. Nagamoto, T., Hattori, Y., Takayama, K., and Maitani, Y. (2004). Novel chitosan particles and chitosan-coated emulsions inducing immune response via intranasal vaccine delivery, *Pharm. Res.*, 21, pp. 671–674.
76. Niidome, T., Yamagata, M., Okamoto, Y., Akiyama, Y., Takahashi, H., Kawano, T., Katayama, Y., and Niidome, Y. (2006). PEG-modified gold nanorods with a stealth character for *in vivo* applications, *J. Controlled Release*, 114, pp. 343–347.
77. Panina-Bordignon, P., Papi, A., Mariani, M., Di, L. P., Casoni, G., Bellettato, C., Buonsanti, C., Miotto, D., Mapp, C., Villa, A., Arrigoni, G., Fabbri, L. M., and Sinigaglia, F. (2001). The C-C chemokine receptors CCR4 and CCR8 identify airway T cells of allergen-challenged atopic asthmatics, *J. Clin. Invest*, 107, pp. 1357–1364.
78. Patil, R., Portilla-Arias, J., Ding, H., Inoue, S., Konda, B., Hu, J., Wawrowsky, K. A., Shin, P. K., Black, K. L., Holler, E., and Ljubimova, J. Y. (2010). Temozolomide delivery to tumor cells by a multifunctional nano vehicle based on poly(beta-L-malic acid), *Pharm. Res.*, 27(11), pp. 2317–2329.
79. Peer, D., Karp, J. M., Hong, S., Farokhzad, O. C., Margalit, R., and Langer, R. (2007). Nanocarriers as an emerging platform for cancer therapy, *Nat. Nanotechnol.*, 2, pp. 751–760.
80. Proudfoot, A. E. (2002). Chemokine receptors: multifaceted therapeutic targets, *Nat. Rev. Immunol.*, 2, pp. 106–115.
81. Roberts, R., Makey, D. G., and Seal, U. S. (1966). Human transferrin. Molecular weight and sedimentation properties, *J. Biol. Chem.*, 241, pp. 4907–4913.
82. Robinson, D., Hamid, Q., Bentley, A., Ying, S., Kay, A. B., and Durham, S. R. (1993). Activation of CD4+ T cells, increased TH2-type cytokine mRNA expression, and eosinophil recruitment in bronchoalveolar lavage after allergen inhalation challenge in patients with atopic asthma, *J. Allergy Clin. Immunol.*, 92, pp. 313–324.

83. Robinson, D. S. (2009). Regulatory T cells and asthma, *Clin. Exp. Allergy*, 39, pp. 1314–1323.

84. Robinson, D. S., Hamid, Q., Ying, S., Tsicopoulos, A., Barkans, J., Bentley, A. M., Corrigan, C., Durham, S. R., and Kay, A. B. (1992). Predominant TH2-like bronchoalveolar T-lymphocyte population in atopic asthma, *N. Engl. J. Med.*, 326, pp. 298–304.

85. Sahoo, S. K., Ma, W., and Labhasetwar, V. (2004). Efficacy of transferrin-conjugated paclitaxel-loaded nanoparticles in a murine model of prostate cancer, *Int. J. Cancer*, 112, pp. 335–340.

86. Sakai, T., Kohno, H., Ishihara, T., Higaki, M., Saito, S., Matsushima, M., Mizushima, Y., and Kitahara, K. (2006). Treatment of experimental autoimmune uveoretinitis with poly(lactic acid) nanoparticles encapsulating betamethasone phosphate, *Exp. Eye Res.*, 82, pp. 657–663.

87. Sallusto, F., Kremmer, E., Palermo, B., Hoy, A., Ponath, P., Qin, S., Forster, R., Lipp, M., and Lanzavecchia, A. (1999). Switch in chemokine receptor expression upon TCR stimulation reveals novel homing potential for recently activated T cells, *Eur. J. Immunol.*, 29, pp. 2037–2045.

88. Sallusto, F., Lenig, D., Mackay, C. R., and Lanzavecchia, A. (1998). Flexible programs of chemokine receptor expression on human polarized T helper 1 and 2 lymphocytes, *J. Exp. Med.*, 187, pp. 875–883.

89. Santeliz, J. V., Van, N. G., Traquina, P., Larsen, E., and Wills-Karp, M. (2002). Amb a 1-linked CpG oligodeoxynucleotides reverse established airway hyperresponsiveness in a murine model of asthma, *J. Allergy Clin. Immunol.*, 109, pp. 455–462.

90. Sehn, L. H., Donaldson, J., Chhanabhai, M., Fitzgerald, C., Gill, K., Klasa, R., MacPherson, N., O'Reilly, S., Spinelli, J. J., Sutherland, J., Wilson, K. S., Gascoyne, R. D., and Connors, J. M. (2005). Introduction of combined CHOP plus rituximab therapy dramatically improved outcome of diffuse large B-cell lymphoma in British Columbia, *J. Clin. Oncol.*, 23, pp. 5027–5033.

91. Serebrisky, D., Teper, A. A., Huang, C. K., Lee, S. Y., Zhang, T. F., Schofield, B. H., Kattan, M., Sampson, H. A., and Li, X. M. (2000). CpG oligodeoxynucleotides can reverse Th2-associated allergic airway responses and alter the B7.1/B7.2 expression in a murine model of asthma, *J. Immunol.*, 165, pp. 5906–5912.

92. Slutter, B., and Jiskoot, W. (2010). Dual role of CpG as immune modulator and physical crosslinker in ovalbumin loaded N-trimethyl chitosan (TMC) nanoparticles for nasal vaccination, *J. Controlled Release*, 148(1), pp. 117–121.

93. Strachan, D. P. (1989). Hay fever, hygiene, and household size, *BMJ*, 299, pp. 1259–1260.

94. Sykulev, Y., Brunmark, A., Jackson, M., Cohen, R. J., Peterson, P. A., and Eisen, H. N. (1994). Kinetics and affinity of reactions between an antigen-specific T cell receptor and peptide-MHC complexes, *Immunity*, 1, pp. 15–22.

95. Umetsu, D. T., McIntire, J. J., Akbari, O., Macaubas, C., and DeKruyff, R. H. (2002). Asthma: an epidemic of dysregulated immunity, *Nat. Immunol.*, 3, pp. 715–720.

96. Van Oosterhout, A. J., Fattah, D., Van, A. I., Hofman, G., Buckley, T. L., and Nijkamp, F. P. (1995). Eosinophil infiltration precedes development of airway hyperreactivity and mucosal exudation after intranasal administration of interleukin-5 to mice, *J. Allergy Clin. Immunol.*, 96, pp. 104–112.

97. Vogel, C. L., Cobleigh, M. A., Tripathy, D., Gutheil, J. C., Harris, L. N., Fehrenbacher, L., Slamon, D. J., Murphy, M., Novotny, W. F., Burchmore, M., Shak, S., Stewart, S. J., and Press, M. (2002). Efficacy and safety of trastuzumab as a single agent in first-line treatment of HER2-overexpressing metastatic breast cancer, *J. Clin. Oncol.*, 20, pp. 719–726.

98. von Maltzahn, G., Park, J. H., Agrawal, A., Bandaru, N. K., Das, S. K., Sailor, M. J., and Bhatia, S. N. (2009). Computationally guided photothermal tumor therapy using long-circulating gold nanorod antennas, *Cancer Res.*, 69, pp. 3892–3900.

99. Wakashin, H., Hirose, K., Maezawa, Y., Kagami, S., Suto, A., Watanabe, N., Saito, Y., Hatano, M., Tokuhisa, T., Iwakura, Y., Puccetti, P., Iwamoto, I., and Nakajima, H. (2008). IL-23 and Th17 cells enhance Th2-cell-mediated eosinophilic airway inflammation in mice, *Am. J. Respir. Crit. Care Med.*, 178, pp. 1023–1032.

100. Walker, C., Bode, E., Boer, L., Hansel, T. T., Blaser, K., and Virchow, J. C., Jr. (1992). Allergic and nonallergic asthmatics have distinct patterns of T-cell activation and cytokine production in peripheral blood and bronchoalveolar lavage, *Am. Rev. Respir. Dis.*, 146, pp. 109–115.

101. Walter, D. M., McIntire, J. J., Berry, G., McKenzie, A. N., Donaldson, D. D., DeKruyff, R. H., and Umetsu, D. T. (2001). Critical role for IL-13 in the development of allergen-induced airway hyperreactivity, *J. Immunol.*, 167, pp. 4668–4675.

102. White, R. R., Sullenger, B. A., and Rusconi, C. P. (2000). Developing aptamers into therapeutics, *J. Clin. Invest*, 106, pp. 929–934.

103. Widera, A., Norouziyan, F., and Shen, W. C. (2003). Mechanisms of TfR-mediated transcytosis and sorting in epithelial cells and applications toward drug delivery, *Adv. Drug Deliv. Rev.*, 55, pp. 1439–1466.

104. Wills-Karp, M., Luyimbazi, J., Xu, X., Schofield, B., Neben, T. Y., Karp, C. L., and Donaldson, D. D. (1998). Interleukin-13: central mediator of allergic asthma, *Science*, 282, pp. 2258–2261.

105. Zhu, Z., Homer, R. J., Wang, Z., Chen, Q., Geba, G. P., Wang, J., Zhang, Y., and Elias, J. A. (1999). Pulmonary expression of interleukin-13 causes inflammation, mucus hypersecretion, subepithelial fibrosis, physiologic abnormalities, and eotaxin production, *J. Clin. Invest*, 103, pp. 779–788.

106. Zingoni, A., Soto, H., Hedrick, J. A., Stoppacciaro, A., Storlazzi, C. T., Sinigaglia, F., D'Ambrosio, D., O'Garra, A., Robinson, D., Rocchi, M., Santoni, A., Zlotnik, A., and Napolitano, M. (1998). The chemokine receptor CCR8 is preferentially expressed in Th2 but not Th1 cells, *J. Immunol.*, 161, pp. 547–551.

Chapter 6

Multifunctional Chitosan Nanocarriers for Respiratory Disease Gene Therapy

Shyam S. Mohapatra,[a,b,d,*] **Subhra Mohapatra,**[c,d] **Gary Hellermann,**[a,b] **and Rhonda R. Wilbur**[b,c]

[a]*Department of Internal Medicine,* [b]*Division of Translational Medicine and Nanomedicine Research Center,* [c]*Department of Molecular Medicine, University of South Florida College of Medicine and* [d]*James A. Haley Veterans Hospital, Tampa, FL 33612, USA*

*smohapat@health.usf.edu

6.1 Introduction

A vast array of technological advances in both nanotechnology and nanoscience has been achieved in the last decade, including advances in drug targeting and delivery applications. State-of-the-art therapeutics, irrespective of their chemical entities, rely on precision-targeting strategies. Diverse arrays of well-known and safe polymeric biomaterials, including chitosan, have been incorporated into multifunctional nanoparticles, which carry not only the drug payload but also targeting and imaging moieties. A myriad of specific cell–targeting approaches are being explored, including antibodies, small molecular ligands, peptides, and other biologics. Examples of successful applications combining genomics and nanotechnology for respiratory disease and cancer are discussed in this review,

Pulmonary Nanomedicine: Diagnostics, Imaging, and Therapeutics
Edited by Neeraj Vij

ISBN 978-981-4316-48-4 (Hardback), 978-981-4364-14-0 (eBook)
www.panstanford.com

with a plasmids encoding nine different antigens of RSV effectively protected mice from viral infection [36]. Chitosan nanoparticles encapsulating the IFN-γ gene also are capable of reversing ongoing asthma in a mouse model [3]. Taken together, chitosan possesses all the attributes required for an ideal *in vivo* drug/gene carrier.

Ehrlich's first magic bullet, arsphenamine, which was discovered in 1909, provided the only cure for syphilis at that time. Treatment involved attaching toxins to antibodies and creating a type of biological "cruise missile" in which the antibody carried and targeted its deadly freight specifically to the site of the invading parasite. The past 100 years of biochemical research have produced a myriad of tools for the design and synthesis of small organic molecules capable of functioning as precision-targeting therapeutics.

Bioscience and medicine have changed unimaginably since Ehrlich's era through sequencing of the entire human genome, knowledge of disease-related genes, discovery of small interfering RNAs (siRNAs) and micro RNAs (mRNAs), and increased awareness of the importance of single nucleotide polymorphisms (SNPs) and posttranslational modifications. These advances have launched promising new directions in research ranging from targeted gene therapy to personalized medicine. We now stand on new frontiers in genomics and nanoscale synthesis, and we anticipate that continued progress in both of these arenas will provide additional opportunities, expanding benefits beyond their original promise into the clinic. Breakthroughs in nanoscale synthesis of functional drug delivery vehicles are expected to have a broad, long-term impact on human health, especially in chemotherapeutics where specificity is desperately needed.

The three main features of nanomedicine are specific detection, targeted delivery, and simultaneous tracking of nanoparticles with indications of the disease status. The nanoparticles we are presenting are multifunctional (also called multifunctional nanoparticles) and recently have been referred to as "theranostics." They compound functional qualities, enabling the "detect, track, and destroy" concepts of early diagnostics, therapy, and follow-up. Theranostics are expected to significantly advance nanotechnology development. Appropriate contrast agents for imaging of a single cell (detection), delivery of therapeutic drugs (destruction), and monitoring of the therapeutic effect (tracking) are key issues in the development of personalized medical care.

A number of biomaterials have been incorporated in the development of nanoparticles that are used for the delivery of drugs and genes to cells and mammals, including humans. These include lipids and polymers such as poly(lactic-co-glycolic acid) (PLGA), polyethylene glycol (PEG), poly ethyleneimine (PEI), cyclodextrin, and chitosan. Of specific relevance to this review are the chitin-based cationic polymers, known as chitosan, which have been used historically as dietary supplements. The last decade has brought multiple advances in the use of chitosan nanoparticles, ranging from their use in drug delivery to the creation of theranostic particles, which are reviewed in this chapter in order to highlight their application in respiratory disease.

6.1.1 What Are Chitosan Nanoparticles?

Chitosan (2-amino-2deoxy-(1→4)-β-D-glucopyranan) is a poly-aminosaccharide normally obtained by the alkaline deacetylation of chitin. It is the principal component of living organisms such as fungi and crustacea (Fig. 6.1). The degree of *N*-acetylation (DA) together with the molecular weight are the most important parameters in its characterization. The DA, which is by definition the molar fraction of *N*-acetylated units, is a structural parameter influencing charge density, crystallinity, and solubility, including the propensity to enzymatic degradation, with higher DAs leading to faster biodegradation rates.

The major physicochemical properties of chitosan include the following:

- Chitosan is a linear polymer of mainly anhydroglucosamine, which behaves as a linear polyelectrolyte at acidic pH.
- Chitosan is nontoxic and bioabsorbable.
- At pH below 6.5, chitosan in solution carries a high positive charge density (one charge per glucosamine unit).
- Since chitosan is one of the few cationic polyelectrolytes, it is an exception to the current industrial high molecular weight polysaccharides, which are mostly neutral or polyanionic.

Chitosan has been sold as a dietary pill for decades. It has been shown to have extraordinary therapeutic properties, and chitosan nanoparticles have great potential to be a safe and effective carrier for genes and drugs.

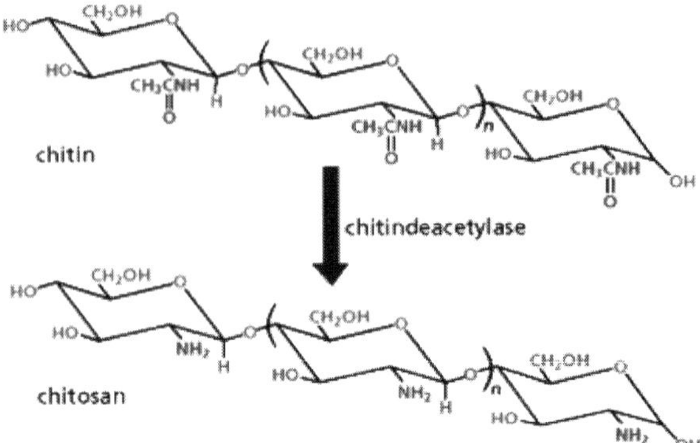

Figure 6.1 The structure of a chitosan polymer. Chitosan (2-amino-2deoxy-(1→4)-β-D-glucopyranan) is a polyaminosaccharide normally obtained by the alkaline deacetylation of chitin. See also Color Insert.

Chitosan has been shown to have many therapeutic benefits. It has strong immunostimulatory properties [1], low immunogenicity, anticoagulant activity, wound-healing properties [2], and anti-microbial properties. Chitosan is nontoxic, nonhemolytic, and slowly biodegradable and has been used widely in successfully controlled drug delivery [3–8]. Moreover, as a carrier, it can provide heat stability to encapsulated or adsorbed vaccines. Chitosan also increases transcellular and paracellular transport across the mucosal epithelium [9], and thus it may facilitate mucosal allergen delivery and modulate immunity of the mucosal and bronchus-associated lymphoid tissue.

Two physicochemical properties are important in chitosan use as a drug and gene delivery agent — the surface charge and solubility. Chitosan is a polysaccharide with cationic ($-NH_3^+$) and alcoholic ($-OH$) groups distributed along the hydrocarbon structure. It exhibits low surface activity due to the absence of large hydrophobic groups in the structure [10]. The surface tension of chitosan in water was found to be very close to the surface tension of pure water at the same temperature, indicating the absence of surface activity as the chitosan molecules are excluded from the air/solution interface [11]. A lower pH results in a decrease in the surface tension presumably due to the protonation of amino groups and increase in hydro-philicity. The reverse occurs at higher pH, at which the hydrophobic character of chitosan increases due to deprotonation of amino acids. Further, it has been reported that the water surface tension decreases linearly with increasing logarithm of chitosan concentration and then levels off [10]. Thus, the reversible surface charge endows chitosan with its "intelligent" drug releasability feature.

Chitosan is readily soluble in dilute acidic solutions with pH below 6. This occurs due to the presence of primary amino groups, which make it a strong base with a pK_a value of 6.3. The degree of ionization and other properties significantly depend on the pH [12]. At low pH values, chitosan is soluble in water due to protonation of the amino groups and presence of a net positive charge. At pH values above 6, the amino groups become deprotonated and chitosan is insoluble due to loss of charge. The solubility of chitosan depends on the degree of deacetylation, and the method of deacetylation used as the pK_a value depends on the degree of *N*-acetylation [13]. Ionic concentration is also a factor determining the solubility of chitosan, and a salting-out effect was observed in excess HCl [14]. Chitosan is soluble in organic acids such as formic, acetic, and lactic acids as it has the ability to form quaternary nitrogen salts in the low pH range [12, 15, 16]. Several critical factors, including temperature and time and degree of deacetylation, particle size, alkali concentration, the ratio of chitin to the alkali solution, and molecular weight play a role in chitosan solubility and hence are extremely relevant to gene/drug delivery using chitosan as a carrier [17, 18].

6.2 Therapeutic Effects and Safety of Chitosan in Human Disease

Chitosan has been reported to be one of the safest polymers studied to date [19]. The Environmental Protection Agency (EPA) has ruled chitosan exempt from its tolerance guidelines because of its nontoxicity, which is evidenced by the scientific literature when searching for chitin, chitosan, *N*-acetyl-D-glucosamine, and D-glucosamine toxicity in humans in the databases Pub Med, Hazardous Substances Data Bank, Integrated Risk Information System, Gene-Tox, Environmental Mutagen Information Center, Toxic Release Inventory, the FDA, the USDA, and ChemIDplus. Animal-feeding studies in which up to 5% of the diet was chitosan failed to show any adverse effects. Despite years of chitosan use in food and nutritional supplements, there are no reported complaints of toxicity reported in these databases.

Chitosan has been tested as an independent drug in several clinical trials for its effect in weight-loss, reduction of cholesterol, and wound healing. Several randomized clinical trials and meta-analyses are summarized with respect to conditions outlined in Table 6.1. No trial to date has measured the effect of chitosan on mortality or morbidity. In general, chitosan has been clinically well tolerated, and none of these clinical trials have shown any serious adverse events. Mild and transitory nausea and constipation have been reported in 2.6–5.4% of the subjects. Using low-molecular-weight chitosan, Jaffers and Sampalis [20] noted 29 predominantly mild adverse events reported by 24 (23%) patients related to the study treatment, most frequently constipation (3.0%) and diarrhea (3.0%).

To evaluate the effect of chitosan on serum cytokine levels in elderly adults, 5.1 g/day of chitosan was administrated to volunteers (age range approximately 74 to 86 years; mean 80 +/– 3 years old) for eight weeks [21]. The clinical study showed that IL-2, IL-12, and TNF-alpha production was minimally increased in the chitosan-administered group compared to the control group; however, there were no statistically significant differences. In the safety study with blood biochemical testing, it has been shown that all safety parameters in the liver remained within normal ranges. There were no significant changes in the values of the electrolytes, blood lipids profiles, glucose levels, or leukocyte numbers.

Concurring with these results, we did not find any safety issues following the administration of chitosan during an eight-week study. This study demonstrated an immune-enhancing tendency at the experimental dose generally used. Finally, in a hemostasis study, Valentine *et al.* [22, 23] showed that there was no significant difference between chitosan–dextran gel and control with respect to crusting, mucosal edema, infection, or granulation tissue formation. Taken together, these studies demonstrate that chitosan as an independent drug is capable of decreasing body weight and decreasing total cholesterol (Table 6.1). A majority of these studies have used chitosan either orally or topically for wound application. Studies also have revealed the safety of chitosan in multiple clinical trials. In addition to therapeutic effects of chitosan by itself, it has been used successfully as a drug carrier in a number of human studies (Table 6.2).

Table 6.1 Therapeutic effects of chitosan

Therapy	Author	Study design	Inference	Remarks
Weight Loss	Jull *et al.* [24]	Meta-analysis: randomized controlled trials. Four weeks minimum. 1,219 overweight or obese adults.	Decrease in total cholesterol. Decrease in systolic and diastolic blood pressure.	High-quality trials indicate that the effect of chitosan on body weight is minimal.
"	Pittler and Ernst [25]	RDBPC trial. 30 subjects received chitosan or placebo daily for 28 days. Measurements were taken at baseline, after 14 days, and then after 28 days.	No significant change in BMI, serum cholesterol, triglycerides, vitamins A, D, E, and beta carotene after four weeks (in the chitosan group).	Chitosan does not reduce body weight in overweight subjects.
Cholesterol Reduction	Wuolijoki *et al.* [26]	RDBPC trial (n = 51). Tested MCCh.	MCCh decreased LDL cholesterol but not HDL cholesterol. Increase in serum triglycerides.	Chitosan is well tolerated. There are no SAE.
"	Bokura and Kobayashi [27]	RDBPC trial (n = 84), female (34–70 yo), 1.2 g/d vs placebo.	Chitosan significantly (p = 0.04) reduced total cholesterol and LDL cholesterol.	
"	Ausar *et al.* [28]	Controlled trial (n = 18). Dyslipidemic type 2 diabetes.	Chitosan decreased LDL cholesterol and increased HDL cholesterol with no change in triglycerides.	There were no SAE; but 29 mild adverse events, including constipation (3%) and diarrhea (3%) occurred.

(*Cont'd*)

Table 6.1 (*Cont'd*)

Therapy	Author	Study design	Inference	Remarks
"	Jaffer and Sampalis [20]	Tested low molecular weight chitosan HEP-40 in RDBPC multicenter trial, $n = 105$, Rx-naïve, 12-week regimen, 3-arm study.	Overall Rx effect was significant. Placebo vs 2,400 mg chitosan once daily ($p = 0.002$) and placebo vs 400 mg chitosan thrice daily ($p = 0.054$).	There was no significant difference for LDL, HDL, or triglycerides.
"	Baker *et al.* [29]	Meta-analysis of chitosan effects on serum lipids. ($n = 6$, total subjects 416).	Used weighted mean different from baseline. Chitosan lowered the total cholesterol ($p = 0.02$).	

Abbreviations: BMI, body mass index; HDL, high-density lipoprotein; LDL, low-density lipoprotein; LMW, low molecular weight; MCCh, microcrystalline chitosan; RDBPC, randomized, double-blind, placebo-controlled; SAE, serious adverse events.

Table 6.2 Chitosan as a drug carrier in human clinical studies

Disease/author	Study design	Inference	Remarks
Psoriasis/Lakshmi *et al.* [30]	RDBPC study ($n = 10$). Compared topical chitosan-niosomal MTX gel vs MTX gel.	Used repeated human insult patch test. Total severity score reduced.	Niosomal MTX is more efficacious than placebo and MTX alone.
Crohn's disease/Hejazi and Amiji [31]	Outpatients ($n = 11$). Chitosan and ascorbic acid mixture (chitosan, 1.05 g/day) for eight weeks.	Significantly increased fat concentration in feces during treatment.	Oral chitosan and ascorbic acid mixture in patients with Crohn's disease is well-tolerated and increases fecal fat excretion.
Hepatic cancer/Wang *et al.* [32]	Tested cisplatin-chitosan microspheres ($n = 6$).	Plasma concentration of platinum and the AUC of microsphere group were lower compared to free drug solution group.	The platinum content in hepatic tissue was 2.92 times greater with cisplatin microspheres than with free drug.

Disease/ author	Study design	Inference	Remarks
Hepatic cancer/ Cho *et al.* [13]	Tested transarterial injection of chitosan-166holmium for SHC to patients (n = 12) with single tumors <3 cm.	Ten patients showed complete response, and two showed partial response.	No grade III/IV toxicity was observed. No toxic death.
Renal failure/ Jing *et al.* [33]	Crossover design in long-term stable hemodialysis patients (n = 80). One week after treatment with control, half were fed 450 mg of chitosan daily for four weeks.	Effectively reduced the total serum cholesterol, urea, and creatinine. Increased hemoglobin and improved physical strength, appetite, and sleep versus control.	Clinically problematic symptoms were observed. Chitosan may be an effective treatment for renal failure patients.

Abbreviations: AUC, area under the curve; MTX, methotrexate.

6.2.1 Chitosan as Gene Therapy

Over the past several years, the Mohapatra lab has developed modified chitosan nanoparticles as a platform technology for DNA-based therapy. We previously reported on the application of chitosan nanoparticles for generating a vaccine to respiratory syncytial virus (RSV). The new chitosan series Nanogene-042 (NG042) is significantly smaller than the parent chitosan as characterized by atomic force microscopy (Fig. 6.2, *left*). NG042 is a unique low-molecular-weight carrier that is heat stable and possesses the capacity for gene transfer and expression both *in vitro* and *in vivo*. When administered intranasally, these particles transfected lung epithelial cells and also cells in the distal lung through macrophages and dendritic cells (Fig. 6.2, *right*). A significant finding is that reconstitution of lyophilized NG042 or incubation of DNA-complexed NG042 for five days at temperatures in the range of 23°C to 55°C did not alter the gene expression profile. Gene expression through intranasally delivered DNA nanoparticles may be detected in lung and other tissues.

Research from our laboratory has shown that various chitosan-based nanoparticle formulations can effectively transfer pDNAs and express genes in lung cells [3, 34–43]. Particles encapsulating plasmids encoding nine different antigens of RSV effectively protected mice from viral infection [36]. Chitosan nanoparticles encapsulating the IFN-γ gene also are capable of reversing ongoing asthma in a mouse model [3]. Taken together, chitosan possesses all the attributes required for an ideal *in vivo* drug/gene carrier.

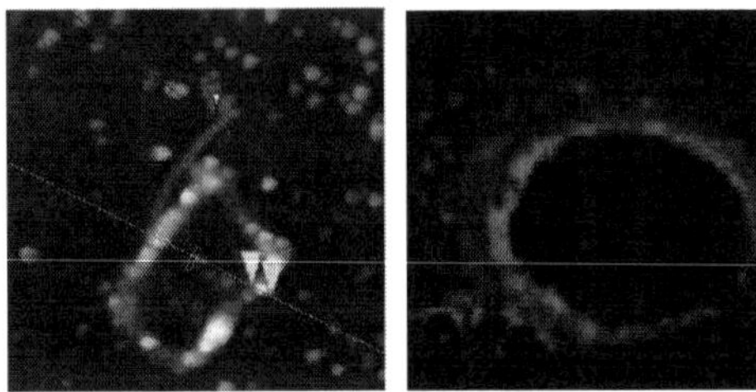

Figure 6.2 Synthesis and characterization of oligochitosan nanoparticles and lung biodistribution. (*left*) Atomic force microscope analysis of nanoparticles shows oligomeric structure complexed with DNA (red arrow). (*right*) NG042 nanoparticles given intranasally to mice deliver genes to both the proximal and distal lung, as shown using the gene for green fluorescent protein (GFP). See also Color Insert.

Chitosan is a biodegradable and biocompatible polymer, and it is useful as a nonviral vector for gene delivery. In order to deliver the pDNA/chitosan complex into macrophages expressing a mannose receptor, Hashimoto *et al.* [44] employed mannose-modified chitosan. The pDNA/mannose-chitosan complexes showed much higher transfection efficiency than pDNA/chitosan complexes in mouse peritoneal macrophages and no cytotoxicity. The observation with a confocal laser microscope suggested differences in the cellular uptake mechanism between pDNA/chitosan complexes and pDNA/man-chitosan complexes. Mannose receptor-mediated gene transfer thus enhances the transfection efficiency of pDNA/chitosan complexes.

To augment the antitumor immune action by *in vivo* IL-12 gene delivery, Kim *et al.* [45] used mannosylated chitosan to induce mannose receptor-mediated endocytosis of IL-12 gene directly into dendritic cells residing within the tumor. Mannosylated chitosan exhibited much enhanced IL-12 gene transfer efficiency to dendritic cells rather than chitosan itself related to induction of murine IL-12 p70 and murine IFN-gamma. In animal studies, the intratumoral injection of MC/plasmid encoding murine IL-12 complex into BALB/c mice bearing CT-26 carcinoma cells clearly suppressed tumor growth and angiogenesis and significantly induced cell cycle arrest and apoptosis. This study provides a new MC-mediated cytokine gene delivery system for cancer immunotherapy.

Research from our laboratory has shown that plasmid DNAs can be encapsulated in chitosan to effectively transfer and express genes in lung cells [46]. Chitosan is also effective in delivering peptides. We have successfully encapsulated the allergen OVA in chitosan. When delivered to mice orally, these particles desensitize the animals (data unpublished).

Roy *et al.* [47] reported an immunoprophylactic strategy using oral allergen-gene immunization to modulate peanut antigen-induced murine anaphylactic response. Plasmid DNA encoding a dominant peanut allergen gene was coupled with chitosan and delivered orally, resulting in transduced gene expression in the intestinal epithelium. Nanoparticle delivery of the allergen resulted in reduced anaphylaxis. This shows oral chitosan–DNA nanoparticles are effective in modulating murine anaphylactic responses and may be useful as a prophylactic for food allergy [47]. Chitosan, therefore, appears to be a good candidate for the safe mucosal delivery of recombinant allergen molecules for immunotherapy. Modifications can be made to chitosan that will enhance its mucoadhesiveness and overall efficacy [37, 42]. The complexes are thermally stable and can be stored at room temperature. They can be reconstituted with water as needed.

6.2.2 Toxicity and Safety of Chitosan Gene Therapy

NG042, consisting of chitosan nanoparticles specially produced for gene therapy, has been extensively studied for toxicity and safety. Blood chemistry analysis of chitosan with plasmid vectors pVAX, pIFN-γ, and pNP-73-102 demonstrated no significant change in physiological parameters, suggesting that it is safe as a gene carrier. In a preclinical toxicology study, the effect of chitosan-pVax vector plasmid and chitosan-IFN-γ nanoparticles given intranasally (100 μg/mouse) four times a week was tested in a group of 100 BALB/c mice (50 males and 50 female). The mice were bled before administration, and after one, two, and three months of administration, 10 different organs were examined. Follow-up and observations were performed for both groups. Each mouse was monitored for body weight and general condition on a daily basis beginning the day prior to the first treatment. The general condition of the mouse included examination of the coat, behavior, and health of the animal. The evaluation of behavior and health of the mouse was determined by observing if the mouse was eating, drinking, and moving without discomfort. At the specified time point (stated above), 10 mice (5 female and 5 male) were euthanized and samples were collected. The samples were evaluated grossly for any signs of toxicity and then the tissue samples were divided in half.

One half of the tissues were fixed and examined histologically to determine the condition of the tissue and cells following the procedure. The other half were frozen and then homogenized. The tissue slurry was evaluated by PCR to determine if the IFN-γ plasmid was present and by ELISA to determine levels of IFN-γ. The cell differential was performed, and blood was analyzed for white blood cell count, red blood cell count, hemoglobin, hematocrit, mean corpuscular volume, mean corpuscular hemoglobin, red cell distribution width, platelet count, mean platelet volume, platelet crit, and platelet distribution width. No significant difference was noted in the values for up to three months after the last administration. An analysis of the sera of naïve mice and mice-administered chitosan and empty plasmid (vehicle) did not show any detectable levels of IFN-γ in any samples (Mohapatra lab, unpublished data).

6.3 Safety and Efficacy Studies in Dogs

Dogs provide an excellent model for allergic rhinitis as they develop allergies similar to humans. Dogs immunized with ragweed (RW) as puppies develop the same allergic immune responses as human allergy sufferers. These allergic dogs show elevated total and serum-specific immunoglobulin E (IgE) with increased numbers of eosinophils in their blood and lungs as well as an increase in airway resistance and a decrease in dynamic compliance after a challenge with RW by inhalation. In a pilot study, the effect of THN108 was investigated on pulmonary resistance and RW-induced eosinophilia using a RW-sensitized beagle model.

To determine the effect of chitosan/IFN-γ pDNA nanoparticles (CIN) on metabolic and immunologic parameters in a pilot study, three RW-sensitized beagles were used for vehicle and compound treatment (in collaboration with Dr. Ted Barrett, Lovelace

Respiratory Research Institute, New Mexico). The beagles were born and maintained in the LRRI dog colony. The dogs were housed in temperature-controlled kennels with access to indoor and outdoor runs.

The dogs were fed 350 g of dry dog food once per day (Wayne Mini Lab Dog Diet 8759, Teklad Premier Laboratory Diets, Madison, Wisconsin), and water was available at all times. The procedure for sensitizing dogs to RW has been described previously [48]. Briefly, beagle puppies were sensitized by subcutaneous injection of 50 mg of RW extract (*Ambrosia artemisifolia*, Greer Labs, Lenoir, NC) combined with 1 mg aluminum hydroxide (Accurate Chemical and Scientific, Westbury, New York) weekly for 4 weeks immediately after birth. The dogs received additional subcutaneous injections every 4 weeks for up to 20 weeks of age. The dogs were then periodically challenged with RW (250 mg deposition in the lung) by inhalation beginning at 24 weeks of age. (Fig. 6.3) Male dogs ranging from 11.5 to 12.5 kg and about six years of age were used for this study.

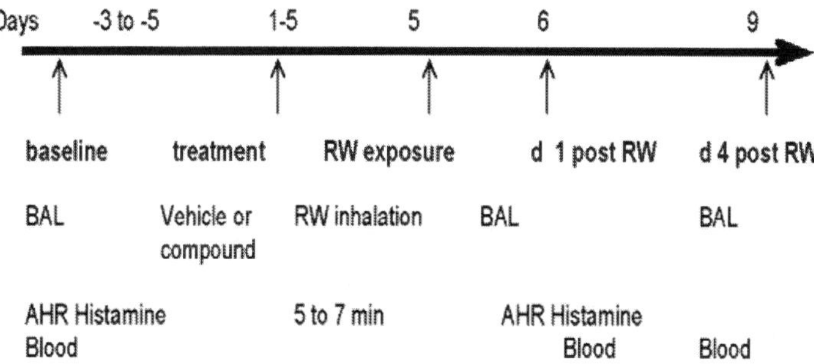

Figure 6.3 A study design for the safety of chitosan gene therapy in dogs. Chitosan in water was given five times (10 ml each) by oral gavage followed by RW challenge 30 minutes after the last treatment. The compound (chitosan nanoparticles adsorbed with canine pIFN-γ) was given 24 hours and 30 minutes (each 5 mg) prior to RW challenge. Airway hyperresponsiveness (AHR) was measured under isoflurane anesthesia. Blood samples for hematology and blood chemistry testing were collected prior to anesthesia at each of the time points. The heart rates and body temperatures were recorded at these time points, and the overall physical appearance of the dogs was evaluated. *BAL*, bronchoalveolar lavage.

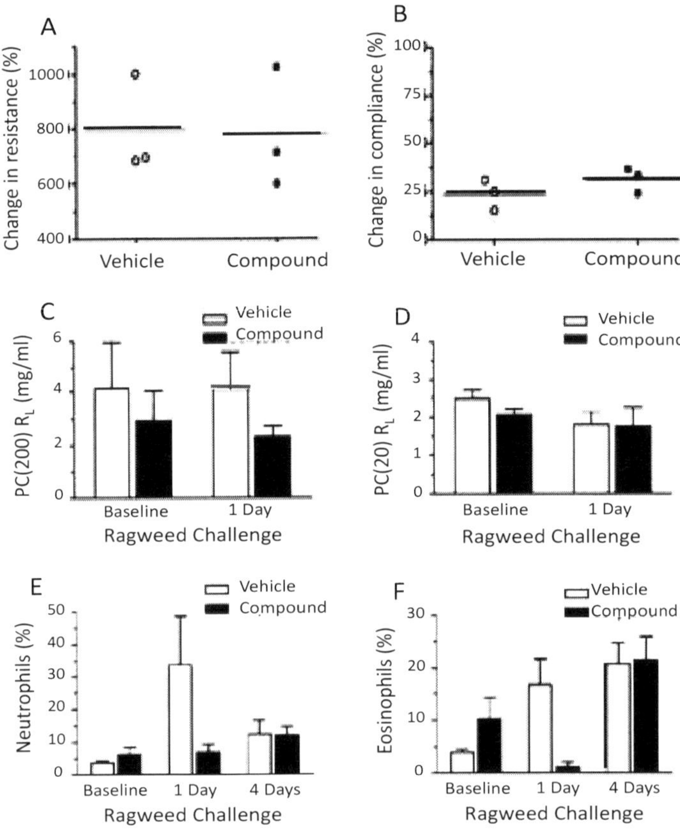

Figure 6.4 Chitosan IFNγ gene therapy is safe in beagle dogs sensitized to ragweed. Dogs were challenged with ragweed to induce allergic lung inflammation, then treated intranasally with chitosan nanoparticles alone ('Vehicle') or nanoparticles containing a plasmid expressing IFNγ ('Compound'). Changes in pulmonary resistance (**A**) and dynamic compliance (**B**) of the lungs in ragweed challenged dogs were measured after treatment and the histamine dose to produce a 200% increase in airway constriction (**C**) or a 20% increase (**D**) was also determined. There was no significant difference in the values at any of the time points. Bronchoalveolar lavage was performed and cell counts of neutrophils (**E**) and eosinophils (**F**) showed that chitosan IFNγ therapy reduced the numbers of infiltrating cells in the lung 24 hours after ragweed challenge.

This study demonstrated that pretreatment with chitosan nanoparticles did not have an effect on the immediate AHR induced by inhalation challenge with ragweed (Fig. 6.4A & B). In addition, there was no difference in bronchoconstriction measured after histamine exposure (nonspecific stimulus) prior to and on day 1 after the ragweed challenge in vehicle-treated dogs compared to those given IFNγ (Fig. 6.4C & D). Blood chemistry data was within normal levels at most time points and did not show significant changes, indicating that nanoparticles did not cause any toxic effects (data not shown). Cell differentials in the blood did not change after the compound treatment compared to the vehicle, although a decrease in eosinophil and neutrophil was seen at day one post-challenge (Fig. 6.4E & F). The overall physical appearance of the dogs was the same before and after the treatment. Chitosan nanoparticles with pIFNγ attenuated infiltration of neutrophils and eosinophils, but not lymphocytes and macrophages, on day 1 after RW challenge compared to vehicle. The vehicle and compound experiments were separated by approximately 10 weeks. Blood samples for hematology and blood chemistry testing were collected prior to anesthesia at each time point. The heart rates and body temperatures were recorded at these time points, and the overall physical appearance of the dogs was evaluated.

6.4 Safety of Chitosan-Gene Nanocomplexes in Nonhuman Primates

We conducted experiments in which we administered either chitosan alone or chitosan-pVax expressing natriuretic peptide receptor A (NPRA) inhibitor to rhesus monkeys via i.p. and oral route. Eight female rhesus monkeys (between 5 and 9 kg) were divided into two groups. Blood was drawn from each animal on day 0 for baseline value to measure the most important parameters for toxicity studies. Animals in the treatment group (n = 4) were i.p. injected with 1 ml (35 mg/ml) chitosan solution once and then fed with 35 mg of chitosan daily for 2 weeks. Control group (n = 4) were i.p. injected with 1 ml of phosphate-buffered saline (PBS) with no chitosan treatment. Additional blood was drawn on days 16, 36, and 66 after chitosan feeding. The analysis of blood chemistry data for each group showed that the treatment by chitosan or chitosan-pDNA did not cause toxicities in the treatment groups (Mohapatra lab, unpublished data).

6.5 Respiratory Disease Applications

Respiratory diseases have been defined historically as either upper or lower lung conditions. Upper lung conditions include both upper airway diseases (such as, hay fever, allergic rhinitis, and rhinosinusitis). On the other hand, lower lung diseases include asthma, chronic obstructive pulmonary disease (COPD), pneumonia, and bronchiolitis, which frequently are caused by respiratory infection. Recently, the concept of "one airway" disease has been suggested, implying that the upper and lower airways share a connected pathology via soluble proinflammatory mediators — the cytokines, chemokines, and effector cells trafficking both upper and lower airways. The treatment of lower lung conditions may be accomplished by addressing either the upper airways (e.g., intranasal) or by treating the lower airway directly.

Due to their large surface area and excellent blood supply, the lungs are an attractive alternative route for drug delivery offering both local and systemic action. By escaping mucociliary or macrophage clearance, inhaled nanopharmaceuticals potentially can be used as a platform for pulmonary sustained release delivery systems. Finally, nanoplexes formed between biodegradable polymeric carriers and DNA/RNA-based drugs can be used to facilitate cellular transfection [17].

Chitosan binds to the mucosal membrane, prolonging the retention time of the formulation contact with nasal mucosa. The time taken for 50% of chitosan microspheres and solution to be cleared from the nasal cavity following nasal administration to human volunteers was evaluated [49]. The control was cleared rapidly with a half-life of 21 minutes. Half-lives of 41 and 84 minutes were recorded for chitosan solution and microspheres, respectively. The retention time was dependent on the cross-linking density of chitosan [50]. Illum *et al.* [51] reviewed their own research using chitosan in solution and powder forms as nasal delivery system for vaccines. They discussed three types of diseases, including nasal influenza, pertussis, diphtheria, and vaccination for these with chitosan powder and solution.

6.5.1 Examples of Chitosan Nanoparticle Applications to Treat Allergic Disease

A growing area of research involves asthma treatment based on plasmid-encoded expression of recombinant therapeutic peptides, such as those obtained from the atrial natriuretic prohormone (ANH) [52]. Difficulties associated with the use of expression plasmids for administering immunomodulatory peptides center principally on the inherent membrane barrier to the passage of charged molecules such as nucleic acids. Naked DNA does not transfect cells well. By complexing the plasmids within a nanoparticle matrix, they are protected from nuclease action. The use of chitosan as a nanoparticle material ensures efficient entry of the complexes into cells. Examples of chitosan nanoparticles with DNA, RNA, drugs, and other peptides are reviewed below.

6.5.1.1 Food allergy and anaphylaxis

Roy *et al.* [47] reported an immunoprophylactic strategy using oral allergen-gene immunization to modulate peanut antigen-induced murine anaphylactic response. The oral administration of DNA nanoparticles synthesized by complexing plasmid DNA with chitosan resulted in transduced gene expression in the intestinal epithelium. Mice receiving nanoparticles containing a dominant peanut allergen gene (pCMVArah2) produced secretory IgA and serum IgG2a. Compared with nonimmunized mice or mice treated with "naked" DNA, mice immunized with nanoparticles showed a substantial reduction in allergen-induced anaphylaxis and reduced levels of IgE, plasma histamine, and vascular leakage. These results demonstrate that oral allergen-gene immunization with chitosan-DNA nanoparticles is effective in modulating murine anaphylactic responses and indicate its prophylactic utility in treating food allergy.

6.5.1.2 Experimental asthma

Since allergic subjects produce relatively low amounts of IFN-γ, a pleiotropic Th-1 cytokine that downregulates Th2-associated airway inflammation and AHR, the effect of CIN was tested for *in situ* production of IFN-γ and its *in vivo* effects [3]. CIN was administered to OVA-sensitized mice in order to investigate the possibility of using gene transfer to modulate ovalbumin (OVA)-induced inflammation and AHR. Mice treated with CIN exhibited significantly lower AHR to methacholine challenge and less lung histopathology. The production of IFN-γ was increased after CIN treatment, while the Th2-cytokines, IL-4 and IL-5, and OVA-specific serum IgE were reduced compared to control mice. AHR and eosinophilia were also significantly reduced by CIN therapy administered therapeutically in mice with established asthma. CIN was found to inhibit epithelial inflammation within six hours of delivery by inducing apoptosis of goblet cells. Experiments performed on STAT4-defective mice did not show reduction in AHR with CIN treatment, thus implicating STAT4 signaling in the mechanism of CIN action. These results demonstrated that mucosal CIN therapy can effectively reduce established allergen-induced airway inflammation and AHR.

Thiolated chitosan nanoparticles (TCNs) were also used to increase the effectiveness of existing drugs using theophylline as a model drug for treatment of asthma [42]. Theophylline is a drug that reduces the inflammatory effects of allergic asthma, but it is difficult to administer at an appropriate dosage without causing adverse side effects. BALB/c mice were sensitized to OVA and OVA challenge to produce an inflammatory allergic condition. They then were treated intranasally with theophylline alone, chitosan nanoparticles alone, or theophylline adsorbed to TCNs. The effects of theophylline on cellular infiltration in BAL fluid, histopathology of lung sections, and apoptosis of lung cells were evaluated to determine the effectiveness of TCNs as a drug-delivery vehicle for theophylline. Theophylline alone exerted a moderate anti-inflammatory effect evidenced by a decrease in eosinophils in BAL fluid, reduction of bronchial damage, inhibition of mucus hypersecretion, and increase in lung cell apoptosis. The effects of theophylline were significantly enhanced when the drug was delivered by TCNs.

6.5.1.3 RSV infection

RSV infection promotes TH2-like immune responses and has been considered a major risk factor for bronchiolitis and asthma. The Mohapatra lab has developed a multigene mucosal vaccination strategy against RSV infection using an RSV cDNA cocktail. [36] This RSV vaccine (NanoRSV-Vax) was produced by cloning nine RSV antigens (NS1, NS2, M, SH, F, M2, N, G, or P in a pVAX plasmid). To administer the vaccine via the mucosal route, the cocktail was formulated with chitosan nanoparticles for delivery via an intranasal route. NanoRSV-Vax has produced results demonstrating that a single dose of about 1 mg/kg body weight is capable of decreasing viral titers by two orders of magnitude (100-fold) following primary infection. This represents a major breakthrough in RSV vaccine development. The immunologic mechanisms for the effectiveness of this prophylaxis include the induction of high levels of both serum IgG and mucosal IgA antibodies, generation of an effective cytotoxic T lymphocyte (CTL) response, and elevated lung-specific production of IFN-γ with antiviral action.

Chitosan nanoparticles are also highly effective as carriers of siRNAs or plasmids encoding siRNA to target cells [53]. The expression of the anti-RSV siRNA in lung epithelial cells has been found to block RSV maturation and inhibit infection in a mouse model of RSV pathology [35]. Mice treated intranasally with siNS1 nanoparticles before or after infection with RSV showed substantially decreased viral titers in the lung as well as decreased inflammation and airway reactivity compared to controls. Thus, siNS1 nanoparticles may provide an effective inhibition of RSV infection in humans. This RNA interference technique is being tested in clinical trials and is currently being evaluated by Alnylam pharmaceuticals [54]. Engineering of specific nanocomplex carriers and siRNA expression plasmids is expected to provide an enhanced antiviral action and minimize off-target effects of siRNA [55].

6.5.2 Nano-Immunotherapy for Allergies

Just as inhaled corticosteroids have become the mainstay of treatment for the symptoms of allergy, subcutaneous allergen-specific immunotherapy (SCIT) is the accepted method for reducing the allergic component of asthma [56, 57]. Conventional allergen-specific immunotherapy (IT) requires a program of yearly injections that are relatively expensive and can have unpleasant side effects. This treatment may be particularly difficult for children. Sublingual IT has been tested extensively in Europe but has not achieved the initial hoped-for success [58]. Oral IT would be an ideal method since it is inexpensive, noninvasive, and free of adverse effects. Current clinical tests of oral IT have shown variable results. More work is clearly needed to develop an effective technique. Research into the application of nanoscale methods for the oral delivery of allergens to suppress overactive allergic response is being studied by several groups, and results have been promising.

Utilizing a standardized mouse model of allergic asthma, Mohapatra and Hellermann [59] showed that chitosan nanoparticles could be used to encapsulate allergen, which could then be delivered orally or as a nasal solution to desensitize the animals. Thus, OVA peptides (Mr 5-20 kDa) were prepared and complexed with chitosan. BALB/c mice ($n = 4$/group) were sensitized by i.p. injection. One group was given chitosan plus OVA peptides (CSOP) by gavage (200 μg), and another was given CSOP intranasally (30 μg). The control group received chitosan alone. Five treatments were given at two-week intervals. One week after the final treatment the animals were challenged with 50 μg OVA given intranasally. After 24 hours, AHR (% Penh) was measured by plethysmography. The airway hyperreactivity to methacholine challenge was lower in mice receiving oral or nasal nanoallergen treatments compared to controls given nanoparticles alone. Differential analyses of BAL fluid showed that lymphocyte, eosinophil, and neutrophil counts were reduced in BAL fluid, overall lung inflammation was reduced in treated mice, and lung sections showed decreased inflammation.

6.6 Future of Immunotherapy

Recombinant allergens and their corresponding cDNAs potentially can be used to prevent allergen-specific IgE responses. As stated previously, the vaccination of mice with a multiepitopic recombinant allergen induced an allergen-specific Ab response that was long lasting despite several booster immunizations [60]. Alternatively, plasmids expressing allergens also could be utilized for prophylactic vaccination. The immunization of rats with a mite allergen-cDNA cloned in a plasmid vehicle resulted in an AL-specific IgG2a and a Th1-like response with no detectable IgE response. This was in marked contrast to immunizing with an allergen that induced an IgE antibody response [61]. These studies suggest that the immunization of the allergen-cDNAs may provide a novel type of prophylactic vaccine against allergic diseases. About a quarter of the population is genetically predisposed toward developing allergic disease. With further advances in gene identification, it may be possible to develop methods to predict atopic predisposition. This, in turn, may allow for vaccination of predisposed individuals against the allergens to prevent the sensitization to which they are exposed.

6.7 Concluding Remarks

The examples of nanoparticle technology application for drug delivery in experimental models already have provided proof of concept; however, how these technologies can best be developed further for clinical studies remains a challenge. It is expected that scaling up the development of nanotechnology methods for drug delivery in allergic diseases and asthma can significantly improve the safety and effectiveness of both existing and new drugs. Unfortunately, predicting the course of nanotechnology in medicine is not possible at this stage of research and development because the "state of the science" only recently has begun to move beyond the simple nanoscale retooling of existing methodologies. Even so, speculative scenarios have ranged from the relatively straightforward improvements in drug delivery and targeting anticipated by the end of this decade to wildly imaginative devices containing multiplex sensors that can identify a specific disease condition such as Th2 cytokines or an increase in reactive oxygen species (ROS), locating the inflammation within the involved tissues via remote-viewing cameras to identify, treat, and correct a given problem. At this time, the role that nanotech applications will play in integrating diagnosis with treatment remains to be established.

Acknowledgments

This work was supported by grants from the NIH (1RO1CA1520050-01), to SSM and SM and a VA Merit Review, Career Scientist Awards, to S. S. Mohapatra. We also like to acknowledge the assistance of Dr. Ed Barrett from Lovelace Respiratory Research Institute in studies involving Beagle dogs.

References

1. Kneuer C, Sameti M, Bakowsky U *et al.* A nonviral DNA delivery system based on surface modified silica-nanoparticles can efficiently transfect cells in vitro. *Bioconjug Chem.* 2000;11(6):926–932.
2. Ezekowitz RA, Williams DJ, Koziel H *et al.* Uptake of Pneumocystis carinii mediated by the macrophage mannose receptor. *Nature.* 1991;351(6322):155–158.

3. Kumar M, Kong X, Behera AK, Hellermann GR, Lockey RF, and Mohapatra SS. Chitosan IFN-gamma-pDNA nanoparticle (CIN) therapy for allergic asthma. *Genet Vaccines Ther.* 2003;1(1):3.
4. Ferkol T, Mularo F, Hilliard J *et al.* Transfer of the human Alpha1-antitrypsin gene into pulmonary macrophages in vivo. *Am J Respir Cell Mol Biol.* 1998;18(5):591–601.
5. Ferkol T, Perales JC, Mularo F, and Hanson RW. Receptor-mediated gene transfer into macrophages. *Proc Natl Acad Sci USA.* 1996;93(1):101–105.
6. Rojanasakul Y, Wang LY, Malanga CJ, Ma JK, and Liaw J. Targeted gene delivery to alveolar macrophages via Fc receptor-mediated endocytosis. *Pharm Res.* 1994;11(12):1731–1736.
7. Wrobel I, and Collins D. Fusion of cationic liposomes with mammalian cells occurs after endocytosis. *Biochim Biophys Acta.* 1995;1235(2):296–304.
8. Behera AK, Kumar M, Lockey RF, and Mohapatra SS. Adenovirus-mediated interferon gamma gene therapy for allergic asthma: involvement of interleukin 12 and STAT4 signaling. *Hum Gene Ther.* 2002;13(14):1697–1709.
9. Kumar M, Behera AK, Matsuse H, Lockey RF, and Mohapatra SS. Intranasal IFN-gamma gene transfer protects BALB/c mice against respiratory syncytial virus infection. *Vaccine.* 1999;18(5–6):558–567.
10. Elsabee MZ, Morsi RE, and Al-Sabagh AM. Surface active properties of chitosan and its derivatives. *Colloids Surf., B.* 2009;74(1):1–16.
11. Pepic I, Filipovic-Grcic J, and Jalsenjak I. Bulk properties of nonionic surfactant and chitosan mixtures. *Colloids Surf., A.* 2009;336(1–3):135–141.
12. Pillai CKS, Paul W, and Sharma CP. Chitin and chitosan polymers: chemistry, solubility and fiber formation. *Prog Polym Sci.* 2009;34(7):641–678.
13. Cho Y-W, Jang J, Park CR, and Ko S-W. Preparation and solubility in acid and water of partially deacetylated chitins. *Biomacromolecules.* 2000;1(4):609–614.
14. Seiichi M, Masaru M, Reikichi I, and Susumu Y. Highly deacetylated chitosan and its properties. 1983;28:1909–1917.
15. Rinaudo M, Pavlov G, and Desbrières J. Influence of acetic acid concentration on the solubilization of chitosan. *Polymer.* 1999;40(25):7029–7032.

16. Ki Myong K, Jeong Hwa S, Sung-Koo K, Curtis LW, and Milford AH. Properties of chitosan films as a function of pH and solvent type. 2006;71:E119–E124.

17. Rinaudo M. Chitin and chitosan - properties and applications. *Prog Polym Sci*. 2006;3(7)603–632.

18. Schiffman JD, and Schauer CL. Cross-linking chitosan nanofibers. *Biomacromolecules*. 2006;8(2):594–601.

19. Rao SB, and Sharma CP. Use of chitosan as a biomaterial: studies on its safety and hemostatic potential. *J Biomed Mater Res.* 1997;34(1):21–28.

20. Jaffer S, and Sampalis JS. Efficacy and safety of chitosan HEP-40 in the management of hypercholesterolemia: a randomized, multicenter, placebo-controlled trial. *Altern Med Rev*. 2007;12(3):265–273.

21. Baek KS, Won EK, and Choung SY. Effects of chitosan on serum cytokine levels in elderly subjects. *Arch Pharm Res*. 2007;30(12):1550–1557.

22. Valentine R, Athanasiadis T, Moratti S, Robinson S, and Wormald PJ. The efficacy of a novel chitosan gel on hemostasis after endoscopic sinus surgery in a sheep model of chronic rhinosinusitis. *Am J Rhinol Allergy*. 2009;23(1):71–75.

23. Valentine R, Athanasiadis T, Moratti S, Hanton L, Robinson S, and Wormald PJ. The efficacy of a novel chitosan gel on hemostasis and wound healing after endoscopic sinus surgery. *Am J Rhinol Allergy*. 2010;24(1):70–75.

24. Jull AB, Ni Mhurchu C, Bennett DA, Dunshea-Mooij CA, and Rodgers A. Chitosan for overweight or obesity. *Cochrane Database Syst Rev*. 2008(3):CD003892.

25. Pittler MH, and Ernst E. Dietary supplements for body-weight reduction: a systematic review. *Am J Clin Nutr*. 2004;79(4):529–536.

26. Wuolijoki E, Hirvela T, and Ylitalo P. Decrease in serum LDL cholesterol with microcrystalline chitosan. *Methods Find Exp Clin Pharmacol*. 1999;21(5):357–361.

27. Bokura H, and Kobayashi S. Chitosan decreases total cholesterol in women: a randomized, double-blind, placebo-controlled trial. *Eur J Clin Nutr*. 2003;57(5):721–725.

28. Ausar SF, Morcillo M, Leon AE *et al*. Improvement of HDL- and LDL-cholesterol levels in diabetic subjects by feeding bread containing chitosan. *J Med Food*. 2003;6(4):397–399.

29. Baker WL, Tercius A, Anglade M, White CM, and Coleman CI. A meta-analysis evaluating the impact of chitosan on serum lipids in hypercholesterolemic patients. *Ann Nutr Metab.* 2009;55(4):368–374.

30. Lakshmi PK, Devi GS, Bhaskaran S, and Sacchidanand S. Niosomal methotrexate gel in the treatment of localized psoriasis: phase I and phase II studies. *Indian J Dermatol Venereol Leprol.* 2007;73(3): 157–161.

31. Hejazi R, and Amiji M. Chitosan-based gastrointestinal delivery systems. *J Controlled Release.* 29 2003;89(2):151–165.

32. Wang YM, Shi TS, Pu YL, Zhu JG, and Zhao YL. [Studies on hepatic arterial embolization with cisplatin-chitosan-microspheres in dogs], article in Chinese. *Yao Xue Xue Bao.* 1995;30(12):891–895.

33. Jing SB, Li L, Ji D, Takiguchi Y, and Yamaguchi T. Effect of chitosan on renal function in patients with chronic renal failure. *J Pharm Pharmacol.* 1997;49(7):721–723.

34. Mohapatra SS. Mucosal gene expression vaccine: a novel vaccine strategy for respiratory syncytial virus. *Pediatr Infect Dis J.* 2003;22 (2 Suppl):S100–103; discussion S103–S104.

35. Zhang W, Yang H, Kong X *et al.* Inhibition of respiratory syncytial virus infection with intranasal siRNA nanoparticles targeting the viral NS1 gene. *Nat Med.* 2005;11(1):56–62.

36. Kumar M, Behera AK, Lockey RF *et al.* Intranasal gene transfer by chitosan-DNA nanospheres protects BALB/c mice against acute respiratory syncytial virus infection. *Hum Gene Ther.* 10 2002;13(12):1415–1425.

37. Lee D, Zhang W, Shirley SA *et al.* Thiolated chitosan/DNA nanocomplexes exhibit enhanced and sustained gene delivery. *Pharm Res.* 2007;24(1):157–167.

38. Wang X, Xu W, Mohapatra S *et al.* Prevention of airway inflammation with topical cream containing imiquimod and small interfering RNA for natriuretic peptide receptor. *Genet Vaccines Ther.* 2008;6:7.

39. Kong X, Wang X, Xu W *et al.* Natriuretic peptide receptor a as a novel anticancer target. *Cancer Res.* 2008;68(1):249–256.

40. Lee D, and Mohapatra SS. Chitosan nanoparticle-mediated gene transfer. *Methods Mol Biol.* 2008;433:127–140.

41. Lee D, Lockey R, and Mohapatra S. Folate receptor-mediated cancer cell specific gene delivery using folic acid-conjugated oligochitosans. *J Nanosci Nanotechnol.* 2006;6(9–10):2860–2866.

42. Lee DW, Shirley SA, Lockey RF, and Mohapatra SS. Thiolated chitosan nanoparticles enhance anti-inflammatory effects of intranasally delivered theophylline. *Respir Res.* 2006;7:112.

43. Kumar A, Sahoo B, Montpetit A, Behera S, Lockey RF, and Mohapatra SS. Development of hyaluronic acid–Fe_2O_3 hybrid magnetic nanoparticles for targeted delivery of peptides. *Nanomedicine.* 2007;3(2):132–137.

44. Hashimoto M, Morimoto M, Saimoto H *et al.* Gene transfer by DNA/mannosylated chitosan complexes into mouse peritoneal macrophages. *Biotechnol Lett.* 2006;28(11):815–821.

45. Kim TH, Jin H, Kim HW, Cho MH, and Cho CS. Mannosylated chitosan nanoparticle-based cytokine gene therapy suppressed cancer growth in BALB/c mice bearing CT-26 carcinoma cells. *Mol Cancer Ther.* 2006;5(7):1723–1732.

46. Pedram A, Razandi M, Kehrl J, and Levin ER. Natriuretic peptides inhibit G protein activation. Mediation through cross-talk between cyclic GMP-dependent protein kinase and regulators of G protein-signaling proteins. *J Biol Chem.* 2000;275(10):7365–7372.

47. Roy K, Mao HQ, Huang SK, and Leong KW. Oral gene delivery with chitosan--DNA nanoparticles generates immunologic protection in a murine model of peanut allergy. *Nat Med.* 1999;5(4):387–391.

48. Glunde K, Guggino SE, Solaiyappan M, Pathak AP, Ichikawa Y, and Bhujwalla ZM. Extracellular acidification alters lysosomal trafficking in human breast cancer cells. *Neoplasia.* 2003;5(6):533–545.

49. Soane RJ, Frier M, Perkins AC, Jones NS, Davis SS, and Illum L. Evaluation of the clearance characteristics of bioadhesive systems in humans. *Int J Pharm.* 1999;178(1):55–65.

50. Genta I *et al. Proceedings of the 23rd International Symposium on Controlled Bioactive Materials* (unpublished), Japan. 1996.

51. Illum L, Jabbal-Gill I, Hinchcliffe M, Fisher AN, and Davis SS. Chitosan as a novel nasal delivery system for vaccines. *Adv Drug Deliv Rev.* 2001;51(1–3):81–96.

52. Hellermann G, Kong X, Gunnarsdottir J *et al.* Mechanism of bronchoprotective effects of a novel natriuretic hormone peptide. *J Allergy Clin Immunol.* 2004;113(1):79–85.

53. Katas H, and Alpar HO. Development and characterisation of chitosan nanoparticles for siRNA delivery. *J Controlled Release.* 2006;115(2): 216–225.

54. DeVincenzo J, Lambkin-Williams R, Wilkinson T *et al.* A randomized, double-blind, placebo-controlled study of an RNAi-based therapy directed against respiratory syncytial virus. *Proc Natl Acad Sci USA.* 2010;107(19):8800–8805.

55. Liu X, Howard KA, Dong M *et al.* The influence of polymeric properties on chitosan/siRNA nanoparticle formulation and gene silencing. *Biomaterials.* 2007;28(6):1280–1288.

56. Nelson HS. Allergen immunotherapy: where is it now? *J Allergy Clin Immunol.* 2007;119(4):769–779.

57. Wallner M, Briza P, Thalhamer J, and Ferreira F. Specific immunotherapy in pollen allergy. *Curr Opin Mol Ther.* 2007;9(2):160–167.

58. Brimnes J, Kildsgaard J, Jacobi H, and Lund K. Sublingual immunotherapy reduces allergic symptoms in a mouse model of rhinitis. *Clin Exp Allergy.* 2007;37(4):488–497.

59. Mohapatra SS, and Hellermann G. (2008) Chitosan nanoparticle-mediated drug delivery: application to allergic disease, in *Advances in Aerobiology, Allergy and Immunology* (ed. Vijay H, and Kurup V), Research Signpost, India.

60. Gordon S, Saupe A, McBurney W, Rades T, and Hook S. Comparison of chitosan nanoparticles and chitosan hydrogels for vaccine delivery. *J Pharm Pharmacol.* 2008;60(12):1591–1600.

61. Hsu CH, Chua KY, Tao MH *et al.* Immunoprophylaxis of allergen-induced immunoglobulin E synthesis and airway hyperresponsiveness in vivo by genetic immunization. *Nat Med.* 1996;2(5):540–544.

Chapter 7

Targeted Delivery to the Pulmonary Endothelium

Yifei Zhang, Jiang Li, Xiang Gao, and Song Li*

Center for Pharmacogenetics, Department of Pharmaceutical Sciences, University of Pittsburgh School of Pharmacy, Pittsburgh, PA 15261, USA

*sol4@pitt.edu

The last decade has seen substantial progress in the development and application of various strategies for the targeted delivery of therapeutic agents to the pulmonary endothelium. However, many problems remain to be solved before these strategies can be used in clinical applications. This chapter describes the *in vivo* and cellular barriers to pulmonary endothelium targeting. Its application in the targeted delivery of various types of therapeutics, particularly protein and gene therapeutics, is also discussed.

7.1 Introduction

Pulmonary drug delivery is an important strategy for the treatment of various human diseases, including lung diseases. Drugs can be delivered to the lung via either the airway or the vascular route depending on the disease under study. Airway delivery has been

Pulmonary Nanomedicine: Diagnostics, Imaging, and Therapeutics
Edited by Neeraj Vij

ISBN 978-981-4316-48-4 (Hardback), 978-981-4364-14-0 (eBook)
www.panstanford.com

employed for the treatment of respiratory tract infections, asthma, and chronic obstructive pulmonary diseases and has an obvious advantage of concentrating the administered agents at the site of action and reducing the undesired systemic effects [1–3]. In addition to lung diseases, aerosol has also been explored as a possible route of administration for the treatment of systemic diseases (e.g., diabetes mellitus) due to the large alveolar surface area, excellent blood perfusion, and a thin alveolar-capillary barrier [3]. Recently, there has also been increased interest in developing strategies for the targeted delivery of various types of imaging and therapeutic agents to pulmonary endothelial cells (ECs), including conventional drugs, radioisotopes, protein therapeutics, therapeutic siRNA, and genes. These studies have important implications as endothelial dysfunction plays an important role in a number of pulmonary diseases, such as pulmonary hypertension, adult respiratory distress syndrome, and metastatic disease to the lung. On the other hand, lung ECs are also an attractive site as the "expression factory" for production and secretion of various types of therapeutic proteins into circulation due to their large surface area. The success of these applications, however, is largely dependent on the development of a vehicle that is capable of efficient delivery of different types of therapeutics with minimal toxicity. The delivery of therapeutic agents to the pulmonary circulation can be achieved via passive or active targeting. This chapter will discuss the pulmonary physiology that affects drug delivery to the pulmonary circulation. Special emphasis will be placed on a discussion of strategies that have been developed for the delivery of different types of therapeutics to the pulmonary circulation.

7.2 Pulmonary Endothelium as a Target for Drug Delivery

7.2.1 Physiological Functions of Lung ECs

The pulmonary endothelium is a thin layer of ECs that lines the luminal surface of blood vessels. The ECs act as a semipermeable barrier between blood and underlying tissue, controlling the passage of materials and the transit of white blood cells into and out of the bloodstream. The vascular endothelium is also a metabolically active tissue that regulates the metabolism and/or generation and release of various types of biologically active substances and, therefore, is critically involved in many physiological processes such as regulation of vascular tone and maintenance of a normal balance between coagulation and fibrinolysis systems [4]. Furthermore, pulmonary ECs are also involved in the immunological and inflammatory responses in the lung [5]

The function of the pulmonary endothelium is subjected to alterations in various disease conditions, including hyperoxia, hypertension, smoke, dust inhalation, pneumonia, cardiac failure, sepsis, thrombosis, diabetes, and cancer [6]. Thus, the pulmonary endothelium is an important target for the treatment of both pulmonary and systemic diseases. The targeted delivery of therapeutics to the pulmonary endothelium could improve the therapeutic effects and minimize the untoward side effects.

7.2.2 Pulmonary Endothelium as a Drug Delivery Target

7.2.2.1 Passive targeting

Intravenous administration is an ideal route for drug targeting to the pulmonary endothelium. This is determined by the physiological anatomy of the pulmonary endothelium. The pulmonary endothelium contains roughly one-third of ECs in the body, which consists of the first large capillary bed that is directly accessible for drugs after intravenous injection [7]. In addition, the lung receives the entire cardiac output of the venous blood. Thus, even a drug carrier that does not specifically interact with the pulmonary endothelium can be engineered to achieve a passive accumulation in the lung [8].

A number of factors affect the efficiency of passive targeting, among which the surface charge and the size of the vector play a major role. As detailed later, the positive surface charge of cationic lipid- or polymer-based delivery systems contributes significantly to the efficient pulmonary gene transfer via the systemic route. In addition to the surface charge, micron-sized particles show preferential and transient accumulation in pulmonary microvasculature after intravenous administration. It has been reported that the optimal size for an efficient passive accumulation of rigid microparticles in rat pulmonary circulation ranges between 6 and 10 μm [9]. However, there is a lack of detailed studies on the effect of particle size and surface charge on pulmonary and systemic hemodynamics.

One major advantage of passive targeting is the simplicity of vector development. Recently, there have been more intensive studies in developing strategies for active targeting to the pulmonary endothelium owing to its better safety profile. Pulmonary ECs express a number of surface antigens that can serve as a target for endothelium-selective delivery, as detailed below.

7.2.2.2 Active targeting via surface antigens

Functional and antigenic heterogeneity of vascular ECs has been well documented along all segments of pulmonary circulation [10, 11]. This provides a basis for developing targeting strategy for cell-type specific delivery. There are many cell surface molecules that have been explored for targeting to the pulmonary endothelium. As a good target determinant, a surface antigen must have an adequate surface density on pulmonary ECs and relatively lower expression on other cell types, particularly those cells in the blood circulation. In general, these EC surface targets can be classified into those that are constitutively expressed and those that are induced under certain physiological or pathophysiological conditions. For example, angiotensin-converting enzyme (ACE) is constitutively expressed on the surface of pulmonary ECs while the expression of intercellular adhesion molecule (ICAM-1, CD54) can be further induced by inflammation and oxidative stress [12, 13]. Table 7.1 summarizes the EC surface markers that have been utilized for EC-specific targeting. Currently, there is a continuous interest in identifying new EC-specific molecules using either proteomics or phage display technology. The latter method allows new ligands to be discovered without the identification and characterization of their corresponding receptors on EC surface.

Most of the studies on EC targeting involve the use of antibodies.

Table 7.1 Endothelial surface markers for targeted delivery to vascular ECs

EC markers	Expression profiles	Physiological functions	Targeted drug delivery systems	Applications
PECAM-1	Ubiquitous EC marker; predominantly localized in inter-EC borders [14]	Cell-cell recognition; adhesion and trans-EC migration of leukocytes [15]	Immunoconjugates for the delivery of reporter enzyme [16] and antioxidant enzyme [17] ; Cationic polymer for gene delivery [18]	Acute lung injury [17]; lung transplantation [19]

EC markers	Expression profiles	Physiological functions	Targeted drug delivery systems	Applications
ICAM-1	Upregulated at inflamed ECs	Leukocyte adhesion [20]; receptor for pathogens [21]; signaling molecule [22]	Immunoconjugates for the delivery of antithrombotic agent [23] and antioxidant enzyme [24]; polymer carriers for enzyme delivery such as CAT [25] and ASM [26]	Imaging agent [27, 28]; Acoustic microbubbles as a contrast agent for ultrasound imaging [29]
ACE	Constitutively expressed; enriched on pulmonary ECs [7]	Converting Ang I into Ang II; vasoconstrictive; pro-oxidant, prothrombotic, and proinflammatory [30]	Immunoconjugates for intrapulmonary delivery of antioxidant enzymes SOD and CAT [24, 31, 32]; Ad vectors for gene delivery [33]	Acute lung injury; pulmonary hypertension; pulmonary coagulopathy
TM	Constitutively expressed; enriched on pulmonary ECs [7]	Converts thrombin into an antithrombotic and anti-inflammatory enzyme [34]	Anti-TM conjugated with GOX to induce acute oxidative lung injury in mice [35]; NPLL for gene delivery to lung ECs [36]	Generates animal model of lung injury
Selectin	Normally absent on the vascular lumen but induced on the pathological endothelium [20]	Facilitates the adhesion of leukocytes to ECs [37]	E-selectin immunoconjugate to deliver dexamethasone [38]; E-selectin-targeted immunoliposomes [39]; adenovirus vector for gene delivery [40]; P-selectin-targeted microbubbles for imaging [41]	Targeted isotopes for radioimaging of activated ECs [42]; targeted microbubbles as ultrasound contrasting agents [41]
Endothelial VCAM-1	Absent on normal ECs but induced on inflamed ECs [14]	Facilitates the adhesion of leukocytes to ECs [37]	Magnetofluorescent nanoparticles modified with phage display-derived peptides for VCAM-1-targeted imaging [43];	*In vivo* imaging of endothelial markers
Integrin	Elevated in the tumor vasculature	Interacts with matrix protein, angiogenesis	Liposomes [44], micelles, [45] and RGD-modified proteins [46] for targeted delivery to ECs	Imaging [47] and therapy [48] targeted to angiogenic endothelium in cancers

(*Cont'd*)

Table 7.1 (*Cont'd*)

EC markers	Expression profiles	Physiological functions	Targeted drug delivery systems	Applications
APP	Enriched in the lung vasculature	A metallo-protease that hydrolyzes N-terminal Xaa-Pro peptide bonds [49]	Peptide ligands available	Potential for lung-specific delivery of therapeutic and imaging agents [50]
LOX-1	Expressed selectively at low levels on ECs; upregulated in hypertension and atherogenesis [51]	Activates nuclear factor-kappaB; induces the expression of adhesion molecules and endothelial apoptosis	Isotope-labeled antibody for the imaging of vulnerable plaque [52]	Imaging for predicting atheroma at high risk for rupture [52]
AnnA1	Normally enriched intracellularly; specifically expressed on tumor ECs but not ECs in the normal organ [53]	Regulates membrane aggregation, inflammation, phagocytosis, proliferation and apoptosis [54]	^{125}I-labeled monoclonal antibody[50]	Radioimmunos-cintigraphic imaging of solid tumor lesions and radioimmuno-therapy [50, 55]
gp60	Albumin receptor	Transcytosis of albumin [56]	Peptide ligands available	Potential for albumin-mediated drug delivery

Abbreviations: ACE, angiotensin-converting enzyme; Ad, adenoviral; AnnA1, annexin A1; APP, aminopeptidase-P; ASM, acid sphingomyelinase; CAT, catalase; GOX, glucose oxidase; ICAM-1, intercellular adhesion molecule-1; LDL, low-density lipoprotein; LOX-1, lectinlike oxidized LDL receptor; NPLL, N-terminal modified poly(L-lysine); PECAM-1, platelet EC adhesion molecule-1; RGD, arginine-glycine-aspartic acid; TM, thermodomodulin; VCAM-1, vascular adhesion molecule-1.

The major advantages of antibody-based ligands are their specificity and high affinity. However, monoclonal antibodies are less stable and potentially immunogenic and can be expensive to produce. Short peptide sequences are much more stable and less immunogenic. RGD peptides are well-established ligands to target the integrin receptors, particularly those on neovasculatures [57]. One common problem with peptide-based ligands is a relatively low affinity. Several approaches have been developed to solve this problem, including a) the incorporation of multiple copies of a ligand on a nanoparticle to improve the overall affinity via a multivalency effect [58]; b) the development of dimer, trimer, or tetramer of a peptide ligand [59]; and c) the cyclization of a peptide ligand to lock the sequence in its most favorable 3-D configuration for interaction with the receptor [60]. Zhang *et al.* reported that cLABL peptide-decorated poly(dl-lactic-coglycolic acid) nanoparticles were highly effective in targeting ICAM-1 on the surface of ECs and in facilitating the subsequent endocytosis and lysosomal trafficking [61]. Aptamers are single-stranded DNA or RNA sequences and have recently been explored as a new class of ligands for cell-type-specific targeting [62]. In addition to high affinity and exquisite specificity to cognate receptors, aptamers can be readily modified chemically to improve their stability and bioavailability [63].

The choice of a ligand for EC targeting is determined by many factors. In addition to the abundance of the receptor on ECs, the intracellular fate of targeted vector is also an important consideration. For example, a ligand that is capable of mediating transcytosis will have the potential in directing the loaded cargos to other types of cells in addition to pulmonary ECs. A unique ligand-receptor interaction can also bring additional therapeutic benefit to the disease under study, as in the case of ICAM-1-mediated targeting for the treatment of inflammatory diseases. In addition to an enhanced delivery of therapeutic agents to inflamed ECs, the ligands for ICAM-1 can block the interaction of inflammatory cells with ECs, thus offering additional therapeutic effect. We are currently exploring small molecule ACE inhibitors as EC-specific ligands. In addition to low or lack of immunogenicity, these ACE inhibitors are anti-inflammatory and antihypertensive, which represent another new type of ligands for targeted therapy of pulmonary inflammatory or hypertensive disorders.

7.2.3 Physiological Barriers for Intravenous Drug Delivery to the Pulmonary Endothelium

There are many barriers to the successful application of EC targeting in diagnostic or therapeutic studies. The targeted agent needs to first effectively home in on the target cells following systemic administration. For those therapeutics whose molecular targets reside inside the cells, the subsequent step of cellular uptake is also essential. The factors that affect the final outcome of targeted delivery vary with the location of the target cells, the delivery systems to be used, as well as the ultimate goal of the targeted delivery. The following is a summary of a number of important *in vivo* and cellular barriers for targeted drug delivery to pulmonary ECs.

7.2.3.1 *In vivo* barriers

Intravenous delivery to the pulmonary endothelium is relatively easy to achieve compared to tissues in the distant locations due to the fact that lung is the first capillary bed the intravenous-injected materials will encounter and that ECs are readily available for interaction with blood-borne drugs or imaging agents. This is particularly the case if the delivery systems are equipped with a high-affinity targeting ligand. Nevertheless, a number of key issues need to be considered, including the protection of the therapeutic agents (e.g., proteins, plasmid DNA, and siRNA) from attack by the degrading enzymes, nonspecific interaction with blood components, and rapid elimination of the delivery vehicles by the reticuloendothelial system (RES). An early study by Maruyama and colleagues [64] showed that the efficiency of liposomal targeting to lung ECs via an anti-TM antibody was significantly improved when the circulation times of liposomes were prolonged via PEGylation [64]. This is particularly important in targeting to endothelium of small- and medium-sized blood vessels. Due to a turbulent and rapid blood flow in these vessels, especially arterial vessels, the interaction of the intravenous-injected delivery system with the endothelium in these locations is less effective compared to the ECs in capillaries. It is thus critical for the delivery systems to have sufficiently long half-lives in the blood to achieve effective interaction with ECs in these blood vessels. Targeted delivery to the endothelium of small blood vessels is clinically significant as these blood vessels play an important role in the regulation of vascular resistance and they are also the sites prone to the development of some vascular lesions such as atherosclerosis.

In addition to the half-life in the circulation, other factors such as the sizes and geometric shapes of the delivery vehicles are also important for their pulmonary accumulation. Micron-sized particles, particularly those with an elliptical shape, are more favorable for targeting to capillary ECs due to passive accumulation in these areas. This will add to active targeting by enhancing the initial interaction with ECs. However, large-sized particles tend to be rapidly eliminated by the RES. Thus, careful optimization of size is needed for each delivery system in intact animals.

The affinity of the ligand for the receptor and the ligand density on the surface of the delivery system also affect the overall targeting efficiency. An increase in the ligand density can significantly improve the overall binding affinity of the particles in virtue of a multivalency effect. However, substantial surface modification may alter the biophysical property of the particles, which may lead to increased uptake by the RES. Thus, the ligand density needs to be carefully optimized for each delivery system.

The above strategies are largely focused on maximizing the amount of targeted agents that can be delivered to ECs at target sites. These might be sufficient for imaging applications as well as the targeted delivery of a therapeutics whose molecular target resides extracellularly. Examples for the latter application include fibrinolytic enzymes [23] and activated protein C (APC) [65]. However, the molecular targets for most of the drugs under investigation are located inside cells. Therefore, it is critical for the targeted agents to reach inside the ECs via endocytosis. As detailed below, cellular uptake and the subsequent steps of intracellular trafficking are a complex process that constitutes many cellular barriers for the effective delivery of various types of therapeutics to ECs, including lung ECs.

7.2.3.2 Cellular barriers

The cell membrane poses the first physical barrier preventing hydrophilic or macromolecular therapeutics from reaching the inside of cells. Early studies used surface-modified proteins through PEGylation or cationization to trigger efficient cell uptake via nonspecific adsorptive endocytosis [66, 67]. Recently, a class of short cell-penetrating peptides (CPPs) has been developed to facilitate the intracellular delivery of small molecules, proteins, liposomes, and other nanoparticles [68–71]. These peptides are either arginine-enriched sequences or contain hydrophobic residues that provide amphipathic properties. The exact mechanism for cellular uptake is not fully understood at present; however, it is known that the clustered positively charged residues on CPPs bind to the negatively charged proteoglycans on the cell surface, which triggers an energy-dependent internalization process. Depending on the copy number of CPPs on the vehicle and cargo size, the uptake can be either endocytosis dependent or endocytosis independent [72]. The efficient delivery to major tissues/organs of recombinant proteins carrying CPPs has been demonstrated *in vivo* [68]. Zhou and colleagues have reported that a chimeric peptide inhibitor of nicotinamide adenine dinucleotide phosphate [NAD(P)H] oxidase can significantly block O_2^- production and attenuate angiotensin

II-induced elevation of blood pressure in mice [73]. CPP-mediated protein delivery may have high efficacy and low toxicity and is readily achievable via either peptide synthesis or the recombinant DNA technique.

In contrast to CPP-mediated delivery, most other studies on targeted delivery involve the use of a ligand/receptor system that mediates intracellular delivery via a relatively well-characterized process of endocytosis. Endocytosis can be classified into phagocytosis and pinocytosis, which refer to the uptake of large particles and the uptake of fluid and solutes, respectively [74]. Phagocytosis is typically restricted to specialized mammalian cells (e.g., macrophage cells), while pinocytosis can occur in all types of cells, which represents the major endocytic pathway in pulmonary ECs. Several pinocytic pathways have been characterized, including clathrin-mediated endocytosis (CME), caveolae-mediated endocytosis (CavME), macropinocytosis, and clathrin/caveolae-independent endocytosis [75]. They differ in the composition of the coat (if any), in the size of the detached vesicles, and in the fate of the internalized particles [76].

CME is the best characterized pathway so far for endocytosis. CME occurs constitutively in all mammalian cells and carries out the continuous uptake of essential nutrients, such as the cholesterol-laden LDL and iron-laden transferrin (Tfn). CME involves the concentration of high-affinity transmembrane receptors and their bound ligands into "coated pits" on the plasma membrane, which are formed by the assembly of cytosolic coat proteins such as clathrin [77]. CME has been shown to be involved in the intracellular delivery of various types of targeted agents in a number of cell types, including ECs [78, 79]. The study by Sohrab *et al.* [80] showed that CME was involved in the uptake of circulating alpha-1 antitrypsin (A1AT) by lung ECs, which played a role in the protective effect of A1AT against cigarette smoke-induced emphysema. This may suggest a new strategy for targeted delivery to ECs.

Caveolae are cholesterol-rich non-clathrin-coated pits of ~70 nm in diameter that are apparent at the luminal and abluminal surfaces of the endothelium and as free vesicles in the cytoplasm. Caveolae are abundant in ECs, composing about 20% of the cell volume [81]. Caveolae assemble by recruiting caveolin-1 (Cav-1), a structural protein of caveolae, into membrane lipid microdomains enriched with cholesterol and sphingolipids at the luminal surface [82]. Caveolin-1 self-assembles into oligomers and mediates the invagination of the plasma membrane into flask-shaped structures.

Caveolae plays an important role in the uptake and transport of various types of molecules, such as albumin, transferrin, and LDL. Caveolin-1 is also associated with a number of important signaling molecules, and caveolae contribute significantly to the endothelium homeostasis. Due to the abundance of caveolae in ECs and their active role in transcytosis, strategies that are targeted to CavME have been exploited for both the intracellular delivery of therapeutic agents and transendothelial delivery [83–85]. The latter approach has important implications in overcoming blood-brain barrier (BBB) for the delivery of drugs to the central nervous system (CNS).

Caveolae- and clathrin-independent endocytosis is initiated from small membrane microdomains called lipid rafts. These lipid rafts are 40–50 nm in diameter and diffuse freely on the cell surface. The mechanisms that govern caveolae- and clathrin-independent endocytosis remain poorly understood. There are several clathrin- and caveolae-independent pathways that differ in their dependence on dynamin, a large GTPase involved in the fission of vesicles from the plasma [86]. A series of studies from Muzykantov's group have unveiled a novel caveolae- and clathrin-independent pathway in ECs named cell adhesion molecule-mediated endocytosis pathway (CAM-ME) [87, 88]. This pathway is rather unique as it is induced by cross-linking of either adhesion molecules ICAM-1 (CD54) or PECAM-1 (CD31), which is mediated by ligand-coated nanoparticles. The CAM-ME appears to result from a multivalency effect since "monomeric" antibody to PECAM or ICAM failed to activate this pathway. In addition, the induction of internalization is dependent on an appropriate size (100~300 nm) of the particles.

The engagement of a particular pathway in endocytosis is affected by many factors, including the ligand/receptor involved, the cell type, and the size and surface properties of the delivery system. This significantly affects the fate of endocytosed materials. In general, most of endocytosed molecules move from the early to the late endosomal compartment and eventually end up in lysosomes. The interior of the endosomal compartment is maintained to be acidic (pH $\approx$ 6) by a vacuolar H^+ ATPase in the endosomal membrane that pumps H^+ into the lumen from the cytosol. Late endosomes are more acidic than early endosomes. The inhibition of endosomal acidification via pharmacological agents may slow down the rate and kinetics of the lysosomal pathway. In addition to a fate of degradation via the lysosomal pathway, some of the endocytosed receptors may recycle back to their original membrane domain while some receptors transfer specific macromolecules from one extracellular space to another by transcytosis.

It is difficult to predict the intracellular trafficking of a delivery system due to our limited understanding of the complex process in endocytosis. Nevertheless, the intracellular fate of a targeted agent can be modulated via tailor-design of a delivery system and/or the pharmacological intervention to meet the specific goal of a targeted therapy. Most of the endocytotic pathways lead to lysosomal accumulation of delivered cargos, which can be employed for lysosomal targeting such as enzyme replacement therapy for the treatment of lysosomal storage disorders. However, for the majority of therapeutic agents under investigation, their molecular targets reside in the cytosol or nucleus. Therefore, the therapeutic agents need to be released from endosomes to exert their biological functions.

Endosomal release has been a subject of intensive study over the last few decades. Many strategies have been developed, and they vary with the delivery systems involved. Table 7.2 summarizes some of the commonly used approaches to facilitate the cytoplasmic release of the vector from endosomes following their endocytosis. These strategies are largely designed to either increase the fusogenic activity of the vector in the acidic endosomal environment or slow down the process of endosomal maturation. Some of the latter approaches may also help to enhance the permeability of the endosome membrane by increasing the osmotic pressure inside the endosome. More discussion on some of the examples can be found in Section 7.3. It should be noted that most of the pharmacological agents that modulate the intracellular trafficking are cell type nonspecific and are associated with significant toxicity in high doses. These agents may have limited applications for *in vivo* uses.

Table 7.2 Strategies that facilitate endosomal release following endocytosis

Endosomal release agents	Mechanism of action	Application examples
Inactivated viral particles; reconstituted virosomes	Lysis/fusion of the endosomal membrane	Adenofection [89] Sendai virosomes [90]

Endosomal release agents	Mechanism of action	Application examples
Fusion peptides	Lysis/fusion of the endosomal membrane	Influenza fusion peptide GALA [91]
Cationic liposomes	Fusion with the endosome membrane; lipid exchange	DOPE-containing cationic liposomes [92]
Anionic liposomes	Lipid hexagonal phase formation; membrane fusion	DOPE-containing anionic liposomes [93]
pH-sensitive polyacrylic acids	pH-induced membrane lysis	Poly(ethylacrylic acid) and poly (propyl acrylic acid) [94]
Cationic polymers/ hydrogels with weak amine groups	Proton sponge; osmotic pressure build up after the influx of counter ions; swelling	PEI [95] pH-sensitive hydrogel [96]

Abbreviation: DOPE, dioleyl phosphatidyl ethanolamine; PEI, polyethylenimine.

Most of the small-sized agents (e.g., chemotherapeutic agents and oligonucleotides) will become biologically available and "freely" interact with their molecular targets following escape from endosomes. However, there is an additional barrier of nuclear transport for macromolecules exceeding certain molecular weight (MW) threshold, including proteins greater than 40 kDa [97] or DNA over 250 bp in length [98]. In fact, for nondividing cells, the nuclear transport of macromolecules is a critical rate-limiting step. A commonly used strategy is to incorporate a nuclear localization signal either directly to the protein sequence or into the delivery system to facilitate the transport of protein or DNA into the nucleus [98]. Alternatively, if certain DNA sequences that can be recognized by nuclear factors are inserted, the resulting plasmid can be transported into nucleus more effectively [99, 100].

Overall, various factors should be taken into account in designing a delivery system to effectively bypass many *in vivo* and cellular barriers. Many strategies that have been developed to overcome the cellular barriers are learned from the "tricks" that viruses have developed over millions of years in escaping the attack by the host immune system [101]. A better understanding of the endocytotic mechanism may lead to improved strategies for targeted delivery to specific subcellular compartments of the pulmonary endothelium, which is beneficial for both diagnostic and therapeutic applications.

7.3 Targeting the Pulmonary Endothelium for Imaging and Therapeutic Applications

7.3.1 Imaging Applications

The surface molecules on pulmonary ECs not only serve as landmarks for the pulmonary endothelium but also have important physiological functions. For example, ACE on the pulmonary endothelium plays an important role in the control of vascular homeostasis by regulating the conversion of angiotensin I to angiotensin II. Under pathophysiological conditions such as lung inflammation and pulmonary hypertension, the expression levels of some of EC surface molecules are significantly altered. Therefore, studies on the expression levels of some of the EC surface molecules may not only offer insights into their physiological and/or pathophysiological roles but also help to monitor the disease progression and predict the prognosis. Furthermore, information on the expression of different surface molecules may help us tailor design a targeting system to maximize the therapeutic effect. Positron emission tomography (PET) and magnetic resonance imaging (MRI) have been investigated to assess the expression of a number of surface molecules on the pulmonary endothelium. PET imaging with small molecule ACE inhibitors as ligands has been used to map the ACE activity in lungs in both animals and humans [102, 103]. This study is important as ACE inhibitors have been used in the clinic for the treatment of various cardiovascular diseases. Significant variations in individual responses to ACE inhibitors have been observed in the clinic due to genetic differences. Therefore, the mapping of ACE activity via PET study may represent an attractive approach for image-guided individualized therapy. The major advantage of ACE inhibitors over antibodies as ACE-targeting ligands is the rapid elimination of nonbound ligands from circulation and, therefore, the ready achievement of a high signal/noise ratio at the target tissue. Recently, there has been increasing interest in the use of nanoparticles for *in vivo* imaging, including targeted imaging of the pulmonary endothelium [28, 43]. One advantage for the latter approach is that imaging agent and therapeutic molecules can be both incorporated into the nanoparticles so the targeting efficiency and the therapeutic outcome can be assessed and correlated at the same time.

7.3.2 Therapeutic Applications

The targeted delivery of therapeutic agents holds promise for various types of clinical applications. For example, the targeted delivery of chemotherapeutic agents to pulmonary circulation can help to concentrate the drugs in the lungs, which may be beneficial for the management of both primary and metastatic lung cancers. The following section will focus on targeted delivery for the treatment of noncancerous lung diseases, for which both protein- and gene-based therapeutics have been extensively studied. Protein therapeutics has the advantage of rapid onset of action but suffers from a rapid turnout and necessitates frequent dosing. On the other hand, gene therapy can achieve a sustained gene expression of a therapeutic protein over a relatively long period of time but is relatively slow in its action. The therapeutic choice is largely determined by the underlying diseases and the ultimate goal of the targeted therapy. Eventually, protein and gene therapeutics can be combined to achieve a synergistic therapeutic effect.

7.3.2.1 Targeted delivery of protein therapeutics

Proteins are an important class of therapeutics for the treatment of various human diseases, including pulmonary diseases. The airway is an attractive route for the administration of therapeutic agents for the treatment of lung diseases since most of the administered materials will be deposited in the lungs given an appropriate aerosol particle size and inspiratory flow rate. The airway route is also valuable for systemic delivery because of a large surface area in the alveolar space. However, the efficiency of airway delivery may be limited when the lungs become severely inflamed, resulting in the accumulation of mucus as well as inflammatory cells and their debris. This problem might be overcome by pulmonary delivery via the vascular route. A dozen of therapeutic proteins have been approved for clinical use or are under preclinical evaluation for pulmonary delivery, including antioxidant enzymes (CAT) and antithrombotic enzymes (activated protein C, urokinase-type plasminogen activator [uPA], or tissue-type plasminogen activator [tPA]). A number of carriers have been developed for targeted delivery of protein therapeutics to the pulmonary endothelium.

7.3.2.1.1 *Endothelium targeting of proteins by immunoconjugates*

Proteins have various reactive groups for conjugation, and additional reactive groups may be chemically introduced into the proteins using heterobifunctional coupling reagents [104]. The methods of conjugating proteins with other ligands are well established [105]. Advantages of immunoconjugates for protein delivery include high conjugation efficiency, adjustable size of conjugates, and biodegradability.

GOX is an H_2O_2-generating enzyme that causes acute oxidative lung injury by vascular immunotargeting. It was used as a model enzyme to study the factors modulating the effects of endothelial targeting [106]. Both anti-PECAM and anti-TM delivered comparable amounts of GOX to the lungs, while anti-TM/GOX inflicted more severe lung injury and pulmonary thrombosis than anti-PECAM/GOX, likely because of TM inhibition [35].

CAT is an antioxidant enzyme that helps to reduce oxidative lung injury. CAT conjugated with PECAM antibody rapidly accumulated in the lung [17, 19] and effectively protected it against anti-TM/GOX-induced pulmonary oxidative stress [17]. It also improved the functions of transplanted lung grafts on rats by reducing the oxidative stress and ameliorating ischemia-reperfusion injury. In addition, this strategy helped to prolong the acceptable cold ischemia period of lung grafts [19]. The targeted delivery of CAT was also investigated with anti-ACE monoclonal antibody (MAb) 9B9 in a rat model of lung I/R injury [32]. 9B9-CAT-treated animals showed significantly lower degree of lung injury compared with nontreated I/R rats [32]. Vascular immunotargeting by anti-ACE MAb-CAT conjugates was further validated as a strategy to augment the antioxidative defense of the pulmonary endothelium and limit reactive oxygen species-mediated ischemia-reperfusion (I/R) injury of the lung during lung transplantation [107]. The protective effect of the immunoconjugates of CAT with MAb 9B9 or MAb 1A29 (specific for ICAM-1) was also demonstrated in isolated rat lungs [24].

The tPA is an antithrombotic agent. It is a serine protease that works by converting inactive plasminogen to the active protease plasmin, which catalyzes the proteolytic degradation of fibrin in clots and helps to dissolve the clot. The Anti-ICAM/tPA conjugate accumulated in the rat lungs, where it dissolved fibrin microemboli [23]. The anti-ICAM-1 may have dual beneficial effects for pulmonary thromboprophylaxis, including enhanced drug delivery to sites of inflammation and the potential anti-inflammatory effect of blocking ICAM-1 [23].

7.3.2.1.2 *Fusion proteins*

The recombinant fusion of therapeutic proteins with antigen-binding fragment of antibodies was also explored for targeted delivery to the pulmonary endothelium. One major advantage of the fusion protein therapeutics is that it overcomes the heterogeneity of the immunoconjugates discussed above. Also, large amounts of homogeneous therapeutics can be readily generated with the bacterial or mammalian expression system. This approach has been used to develop a targeted uPA for restoring perfusion after not only thrombotic vascular occlusion but also thromboprophylaxis. A single-chain variable fragment (scFv) of a PECAM-1 antibody was fused with low-MW single-chain prourokinase plasminogen activator (lmw-scuPA) to generate the prodrug scFv/lmw-scuPA, which accumulated in the lungs of wild-type C57BL6/J, but not PECAM-1 null mice. Cleavage by plasmin generated fibrinolytically active 2-chain lmw-uPA, which effectively lysed the pulmonary emboli [108]. Furthermore, the native plasmin activation site in scFv/lmw-scuPA was replaced with a thrombin activation site to avoid premature activation to limit systemic toxicity [109]. Future improvements of recombinant fusion proteins include the utilization of a humanized antigen-binding domain and large-scale production of fusion proteins.

7.3.2.1.3 *Immunotargeting via avidin/biotin immunoconjugates*

Despite the demonstrated success with direct antibody-enzyme conjugates as discussed above, subsequent studies have shown this approach only exhibited limited efficacy. One major limitation with this approach is a relatively poor internalization efficiency with the direct immunoconjugates. Interestingly, a number of studies from Muzykantov's group have shown the efficiency of internalization can be significantly improved via the use of the avidin/biotin system.

This involves the individual biotinylation of the endothelium-specific antibody and the therapeutic enzyme. Subsequent mixing of the biotinylated antibody and enzyme with straptavidin (SA) leads to the formation of small-sized particles with both the enzyme and the antibody exposed on the surface. For example, the development of SA/biotinylated anti-PECAM (b-anti-PECAM)/b-CAT ternary complexes markedly stimulated their internalization by ECs *in vitro* as well as their uptake into either perfused rat lungs or lungs of intact animals [110]. These EC-specific antibody/enzyme complexes accumulated in the lungs after intravenous injection and effectively protected the cells against H_2O_2-induced injury *in vitro* and *in vivo* [110]. A similar approach was also used to deliver CAT and superoxide dismutase (SOD) enzymes specifically to the pulmonary endothelium using anti-ACE antibody MAb 9B9 (b-MAb 9B9) as a targeting ligand [31]. To further investigate the mechanism of this strategy, a reporter enzyme beta-galactosidase (β-Gal) was incorporated in SA/b-anti-PECAM/b-β-Gal complexes [16]. The immunoconjugates effectively accumulated in the lungs and induced a marked elevation of β-Gal activity in the lung tissue persisting for up to eight hours after injection. Predominant intracellular localization of anti-PECAM/SA/β-Gal was demonstrated by confocal microscopy and electron microscopy in cultured cells and in the pulmonary endothelium of intact mice [16].

7.3.2.1.4 *Polymer-based nanocarriers for protein delivery*

One major advantage of antibody-protein conjugates is their simplicity. However, like protein therapeutics alone, these conjugates are associated with a number of drawbacks, including rapid clearance, a lack of protection against degrading enzymes, and inadequate concentration at the target site [111]. A polymer-based drug delivery system (DDS) may offer solutions to these problems. For example, biodegradable polymer nanocarriers (PNCs) can encapsulate enzymes such that the particles are permeable to the substrate but impermeable to proteases. The major challenge for PNC-mediated protein delivery is the poor loading efficiency and the instability of the proteins during the formulating process. To prevent the unfolding and inactivation of proteins during the encapsulation procedure, a novel freeze-thaw encapsulation strategy was designed that attained 20% loading efficiency of CAT into PNCs and protected CAT from lysosomal degradation [112]. Recently, Simone *et al.* conjugated the cargo enzyme CAT with PEG to enhance its encapsulation within diblock copolymer nanocarriers [113]. Both filamentous and spherical particles benefited from PEG modification with increased protein-loading efficiency and enhanced resistance to protease degradation.

Muro *et al.* investigated the intrinsic and extrinsic factors that control the targeting of anti-ICAM-PNCs to ECs [114]. Anti-ICAM-PNCs have shown markedly higher affinity to ECs *in vitro* compared to naked anti-ICAM, as well as higher accumulation and specificity to the lung *in vivo*. They concluded that reformatting monomeric anti-ICAM into high-affinity multivalent PNCs boosts the vascular immunotargeting [114].

One unique feature with an ICAM-1-mediated targeting system is that some of the ICAM-1 in the endocytic vesicles can be recycled to the cell surface [115]. The reappearance of ICAM-1 on the cell surface within a short period of time will make repeated dosing possible. Interestingly, the second-dosed particles showed significantly slower kinetics in the lysosomal trafficking [115]. This will increase the chances for the loaded enzymes to escape from the endosomes before they end up in the lysosomes. Such strategy has resulted in a prolonged therapeutic effect [115].

In addition to the ligand effect, EC targeting via polymer particles is also affected by the carrier geometry [116]. The morphology of poly (1-naphthylamine) (PNA) and the loading efficiency can be adjusted by changing the block ratio of amphiphilic copolymers. Increasing of the hydrophobic polymer fraction leads to a transition from spherical particles to filamentous carriers that can encapsulate significant levels of the protease-resistant enzyme [111]. In addition, the MW of PEG-b-PLA diblock copolymers has been proved to control the structural properties of filamentous PNC. The stiffness, length, and thickness of filamentous-PNC increased with increasing MW [117]. It has been shown that the shape of anti-ICAM-1-coupled particles affects not only the initial specific binding but also the rate of endocytosis and lysosomal transport within ECs. Disks showed longer half-lives in circulation and higher targeting specificity in mice whereas spheres were endocytosed more rapidly [116]. In addition, intracellular trafficking was shown to be affected by the particle size. Micron-size carriers showed prolonged residency in prelysosomal compartments, beneficial for endothelial antioxidant protection by the delivered CAT, while submicron carriers trafficked to lysosomes more readily, which is more suitable for acid sphingomyelinase (ASM) enzyme replacement for the treatment of lysosomal storage diseases [116]. Therefore, the rational design of protein carriers by modifying those factors can help optimize their pulmonary endothelium targeting effect.

7.3.2.2 Targeted gene delivery to the pulmonary endothelium

Gene therapy holds great promise for the treatment of various human diseases, including pulmonary disorders. This can be achieved via either overexpression of a functional transgene (gene therapy) or the silencing of a diseased gene whose overexpression is involved in the pathogenesis and/or development of a pulmonary disease (oligonucleotide or siRNA therapy). However, the development of an efficient, specific, and safe gene delivery system remains the bottleneck. Both viral vectors and nonviral vectors have been explored for gene delivery. The advantages of viral vectors include high gene transfer efficiency and possible long-term gene expression; however, there are safety concerns over their clinical uses. On the contrary, nonviral vectors usually have a better safety profile. The last two decades have seen great progress in the development of various types of nonviral methods, among which lipid- and polymer-based delivery systems have been extensively investigated for both *in vitro* and *in vivo* gene delivery.

7.3.2.2.1 *Lipid-based vectors*

Nucleic acid-cationic liposome complexes, also known as lipoplexes, are among the best studied systems for gene delivery to pulmonary ECs via the vascular route. The first report of *in vivo* gene transfer with lipoplexes appeared in 1989 [118], shortly after cationic liposomes became a popular transfection reagent for culture cells [119]. A relatively weak but clearly measurable level of transgene expression in mouse lungs was demonstrated following either intravenous administration or airway instillation [120, 121]. These early works were conducted using a formulation that was optimized from *in vitro* transfection. Subsequent studies from the same group with prostaglandin G/H (PGH) synthase transgene demonstrated an anti-inflammatory and protective effect against LPS-induced vascular injuries in large animal models [122]. The intravenous delivery of lipoplexes [118, 123] or cationic lipid-polycation-

DNA (LPD) complexes (or lipopolyplexes) [124] predominantly transfects pulmonary ECs with an efficiency about two to three orders of magnitude higher than that in other organs [125, 126]. Interestingly, some alveolar epithelial cells were also transfected, suggesting a possibility that lipoplexes are capable of moving across the endothelial lining and transfect the airway epithelial cells [127, 128]. It was subsequently discovered that the level of pulmonary gene transfer can be greatly enhanced by the optimization of several parameters. A high lipid-to-DNA ratio is necessary to promote efficient *in vivo* transfection [124, 126, 129]. The lipid composition also plays a key role in the *in vivo* transfection efficiency [129]. Liposomes made from a pure double-chain cationic lipid, 1,2-dioleoyl-3-(trimethylammonium) propane (DOTAP), or DOTAP-cholesterol were much more efficient than liposomes composed of DOTAP-DOPE, which were optimal for *in vitro* transfection [126, 129, 130]. Furthermore, the physiochemical properties of liposomes such as the particle size could also affect the overall transfection efficacy. A systematic study using liposomes prepared from DOTAP and cholesterol showed that multilamellar liposomes of 400–800 nm were more active than multilamellar liposomes of larger sizes or small unilamellar liposomes prepared by the sonication method [131].

Despite the fact that pulmonary gene transfer can be readily achieved in animal models, pharmacokinetics and safety profiles of cationic lipid vectors need to be improved before their clinical application. The treatment-related toxicity seems to be multifactorial and is characterized by acute as well as delayed responses. A vasoreactive response could occur shortly after an intravenous injection of lipoplex, possibly due to a complex interaction between positively charged lipoplex and blood components and the subsequent interaction with the pulmonary ECs. This is followed by local lung inflammation and hepatotoxicity. Both lipid and DNA components and the overall physical properties of the complexes (such as the size) are believed to contribute to the toxicity. A high level of toxicity is noticed in animals treated with lipoplex containing an excessive amount of liposomes, particularly those with large sizes. A study revealed that small DOTAP-cholesterol liposomes prepared from a substantially hydrated lipid film under elevated temperature, followed by extrusion, gave rise to a large number of unilamellar liposomes with somewhat deformed/elongated morphology. DNA was well encapsulated/protected with these cationic liposomes, and the resulting lipoplexes were 200–400 nm in size. Such lipoplexes transfected pulmonary ECs efficiently at lower lipid-to-DNA ratios compared to those made from multilamellar liposomes and consequently showed reduced toxicity in animals [130]. Another study suggested that improved DNA encapsulation accomplished at a low ionic strength could lead to more potent lipoplexes capable of transfecting pulmonary ECs at reduced dosage and thus reduced toxicity [132]. Efficient *in vivo* gene transfer could also be achieved with a reduced lipid-to-DNA ratio if plasmid DNA was first condensed with a cationic polymer, such as protamine, in the form of LPD [124].

The surface modification of cationic liposomes with a shielding polymer such as PEG significantly increases the circulation times and reduces toxicity but also leads to decreased interaction with target cells. Tfn or an antibody specific to the Tfn receptor has been conjugated to the shielding polymer to provide targeting capacity to these shielded lipoplexes, resulting in an enhanced delivery of DNA and siRNA to the cell populations (such as tumors or endothelial linings in BBB) that over express these receptors [133, 134]. Another solution to the toxicity is to use a conditional cationic lipid whose head group is positively charged only at nonphysiological pH such as 1,2-dioleoyl-3-dimethylammonium propane (DODAP). DOTAP is an ionizable amino-lipid that has a pK_a of around 6. At acidic pH, it is positively charged and can effectively interact with DNA or short oligonucleotides and facilitate their entrapment inside lipidic particles. The subsequent adjustment of the external pH to neutral pH values results in particles with a neutral surface charge, which is more biocompatible with our biological system. Greater than 80% of input oligonucleotides can be encapsulated into these particles, named stabilized antisense-lipid particles (SALP) [135]. We have shown that this system can be employed to achieve the targeted delivery of antisense oligonucleotides or siRNA to pulmonary ECs *in vitro* and *in vivo* following the incorporation of an EC-specific ligand such as anti-TM antibody [136]. Importantly, intravenous injection of this neutral lipidic vector containing oligonucleotides did not affect either (a) pulmonary or systemic hemodynamics or (b) the blood chemistry in catheterized mice [136].

In addition to improving delivery systems, efforts have also been made to reduce the DNA-related immunotoxicity. Unmethylated cytosine-phosphodiester-guanine (CpG) motifs in bacterial plasmid DNA are "danger" signals that trigger proinflammatory cytokine response, which not only contributes to the toxicity but also causes the inhibition of transgene expression [137–139]. Cationic lipids augment the CpG immune response by enhancing the uptake of liposome/DNA complexes by immune cells such as macrophages. Cationic lipids also appear to sensitize the immune cells to the CpG-mediated immune response (Ma Z and Li S, unpublished data). A number of strategies have been developed to counteract the CpG immunotoxicity, including a) the use of short DNA fragments with reduced CpG contents [140], b) the removal of most of the CpG motifs via mutagenesis in a way that does not affect the transgene expression of resulting plasmid [141], and c) the use of anti-inflammatory agents [139]. Interestingly, CpG toxicity can also be reduced via a simple protocol of sequential injection, that is, the injection of free cationic liposomes followed by the injection of "naked" plasmid DNA [142]. This protocol was originally developed to study the mechanism of cationic lipids-mediated pulmonary gene transfer. Surprisingly, this protocol gave a transfection efficiency that is higher than that with preformed lipoplexes. In addition, the sequential injection protocol is associated with significantly reduced proinflammatory cytokine response.

7.3.2.2.2 *Polymer-based carriers*

The first study on vascular gene delivery to the lung was conducted with poly(L-lysine) (PLL), which was coupled to the anti-TM antibody [36]. Similar to the transfection of other cell types with PLL-based systems, only a low level of transgene expression was observed, possibly due to a lack of an effective endosome-releasing mechanism. Interestingly, highly compact unimolecular polylysine/DNA particles showed excellent transfection efficiency *in vivo*. These particles were generated via either forming PLL/DNA complexes at high NaCl concentrations [143] or using chemically well-defined PEGylated PLL [144]. Incorporation of an antibody fragment to the polymeric immunoglobulin receptor permits the transfection of airway epithelial cells following transcytosis across ECs [145]. This strategy offers an alternative approach for gene delivery to airway epithelial cells of inflamed lungs where copious airway mucus constitutes a physical barrier to airway gene delivery.

PEI is a polymer that has been most intensively studied for *in vitro* and *in vivo* gene delivery, including airway and intravenous lung gene delivery [146]. Both the *N/P* ratio and the polymer structure affect gene transfer efficiency of polymer-based vectors. A high level of gene expression was found in the lung following the intravenous administration of plasmid DNA complexed with linear PEI (LPEI), with a low level of gene expression found in the heart, the liver, the spleen, and kidneys. Transgene expression was primarily located in the alveolar region, and both ECs and epithelial cells were transfected [147]. Interestingly, branched PEIs (BPEIs) are significantly less active than LPEI in systemic gene delivery to the lung although BPEIs are effective in airway gene delivery [147]. In addition, BPEIs were associated with significantly higher levels of acute toxicity, including sudden death at high doses [148]. Efficient gene transfer via PEIs is believed to be due to the abundant titratable weak amine groups in their backbone and side chain groups. They may act as a proton sponge that effectively neutralizes the protons in endosomal compartments. This will slow down the transition between endosomes to lysosomes and help to avoid the destructive degradation of plasmid DNA in the lysosomes. In addition, the accumulation of chloride counter-ion that is associated with continuous proton pumping may trigger an influx of water due to an increased osmotic pressure. This may cause endosomes to swell and rupture [95, 146] and allow the release of DNA into cytoplasm. However, the underlying mechanism for the different behavior between LPEI and BPEI in pulmonary gene transfer remains largely unknown.

One major concern with LPEI is its nonbiodegradability, which may be responsible for its *in vivo* toxicity. To develop polymers for efficient gene transfer with minimal toxicity, numerous attempts have been made to either modify the surface properties of PEI molecules or develop new cationic polymers with more biodegradable properties. Various chemical modifications on PEIs aimed at eliminating a portion of the amine groups and introducing sugar or PEG nonionic hydrophilic groups have led to improved transfection efficiency and/or reduced toxicity [149, 150]. Targeting ligands such as antibodies were incorporated to simultaneously provide shielding effect and target selectivity [151]. We have previously shown [18] that the transfection efficiency of PEI to lung ECs can be enhanced via conjugation with an EC-specific ligand (anti-PECAM) antibody. In addition, this leads to a significant decrease in the proinflammatory cytokine response.

Low-MW PEIs have an intrinsic low cytotoxicity compared to PEIs of a high MW. However, these low-MW PEIs also have low transfection efficiency. Cross-linking of low-MW PEI through biodegradable linkages [152] was able to generate PEI analogs with reduced toxicity and efficient transfection potency. Polyaminoesters are biodegradable copolymers that contain many amine groups and degradable β-ester linkages. Works from Leong's [153] and Langer's groups [154] have demonstrated that libraries of such polymers with structure diversity can be readily generated and screened for transfection activities. Several of these compounds have equal or better transfection efficiency with much reduced cytotoxicity [155]. We have recently generated a library of polyhydroxylalkyleneamines with hydroxyl and alkylhydroxyl groups linked to a polyamine backbone (PHA) [156]. These polymers are generated by directly cross-linking several dichloro alkylating agents with selective diamines. Disulfide bonds can be introduced to polymer backbones by copolymerization with cystamine. The PHA polymers have a backbone structure similar to that of LPEI but have adjustable spacing between the amine groups. In addition, PHA polymers were designed to have less charge density, abundant hydroxyl and alkylhydroxyl "soft" groups, and potentially biodegradable disulfide bonds in the backbone. These are structural features that are potentially critical for reducing the toxicity of the polymers. Our initial *in vitro* screening has identified several PHA polymers that are more efficient than LPEI in transfection. More importantly, these polymers are significantly less toxic. Studies are currently underway to examine their potential in pulmonary gene transfer in intact animals.

7.3.2.2.3 *Viral vectors*

The advantages of viral vectors in *in vivo* gene delivery include relatively high efficiency and long duration of gene expression for certain types of viral vectors such as adeno-associated virus (AAV) and lentiviral vector. A number of vectors have been used for pulmonary gene transfer via airway administration such as Ad vectors. Gene expression following intravenous administration of Ad vectors is largely limited to liver. The ineffectiveness of lung gene transfer by viral vectors via the vascular route is mainly due to a difficulty in ensuring significant residence time in the lung and/or paucity of receptors for adenovirus on the endothelium [157]. To solve this problem, Reynolds *et al.* developed a bispecific antibody to both Ad vectors and ACE, which is preferentially expressed on pulmonary ECs. The resulting conjugate efficiently redirected the Ad vectors to the lung with confirmed specificity to ECs [33]. Compared to the untargeted vector, premixing of this conjugate with an Ad vector resulted in a 20-fold increase in both Ad DNA localization and luciferase transgene expression in the lungs. Reduced gene expression in nontarget organs was also observed with over 80% reduction in liver, although the absolute gene expression level in the lung was still substantially lower than that in the liver [33]. The selectivity of transgene expression for lung ECs was further improved when this targeting strategy was combined with the use of an endothelial-specific promoter [158].

EC targeting via viral vectors can also be achieved via genetic modification to render expression of EC-specific ligand on the surface of viral particles. Work *et al.* [159] have reported that the incorporation of EC-specific peptide cDNA into the VP3 region of the AAV-2 capsid led to a significant improvement in the homing of viral particles to the vascular ECs, particularly lung ECs. We have shown that redirecting of Ad vector to pulmonary ECs can be achieved via a simple sequential injection protocol, that is, the injection of cationic liposomes followed by the viral vector [157]. An approximately 40-fold increase in transgene expression was achieved in the lung compared to Ad vector alone. This was associated with a significantly decreased expression level in liver. This protocol was similarly used before to mediate the delivery of plasmid DNA to pulmonary ECs [142]. It is possible that a similar mechanism is involved for gene transfer via plasmid DNA and Ad vector; however, a much lower dose of cationic liposomes is required for the sequential injection of Ad vector. The injection of preformed cationic liposome/Ad complexes resulted in a significant inhibition of gene transfer in both lungs and liver.

7.4 Conclusion

There has been significant progress over the last decade in our efforts toward developing strategies for the targeted delivery of various types of therapeutics to the pulmonary endothelium. Future work will be focused on a number of areas, including (1) a better understanding of *in vivo* and cellular barriers that affect effective EC targeting, (2) further development of more efficient EC-specific ligands, particularly small molecule ligands for the selective delivery of different types of therapeutics, (3) development of new "smart" biomaterials to achieve drug release in a controlled manner, and (4) systematic evaluation of the therapeutic efficacy and toxicity of the targeted therapeutics in animal models of pulmonary diseases, such as lung injury and pulmonary hypertension. This includes studies to evaluate the effect of targeted therapeutics on pulmonary and systemic haemodynamics. It should be mentioned that a number of novel nanodelivery systems have recently been developed and successfully utilized in targeted delivery to various types of tissues, such as liver and tumors [160, 161]. Experience from these studies shall also benefit our efforts in developing strategies for lung EC targeting. The completion of these studies will help to facilitate the development of better strategies for the targeted therapy of pulmonary diseases with minimal toxicity.

Acknowledgments

This work was supported by National Institutes of Health Grants HL-68688 and HL-91828.

References

1. R. Dalby, J. Suman, Inhalation therapy: technological milestones in asthma treatment. *Adv. Drug Deliv. Rev.* 55(7) (2003) 779–791.
2. J.D. Suman, Nasal drug delivery. *Expert Opin. Biol. Ther.* 3(3) (2003) 519–523.
3. G. Scheuch, M.J. Kohlhaeufl, P. Brand, R. Siekmeler, Clinical perspectives on pulmonary systemic and macromolecular delivery. *Adv. Drug Deliv. Rev.* 58(9–10) (2006) 996–1008.
4. M. Simionescu, N. Simionescu, Functions of the endothelial cell surface. *Annu. Rev. Physiol.* 48(1) (1986) 279–293.
5. R. Lucas, A.D. Verin, S.M. Black, J.D. Catravas, Regulators of endothelial and epithelial barrier integrity and function in acute lung injury. *Biochem. Pharmacol.* 77(12) (2009) 1763–1772.
6. V.R. Muzykantov, V.P. Torchilin, *Biomedical Aspects of Drug Targeting*. Kluwer Academic Publishers, 2002.

7. S.M. Danilov, V.D. Gavrilyuk, F.E. Franke, K. Pauls, D.W. Harshaw, T.D. McDonald, D.J. Miletich, V.R. Muzykantov, Lung uptake of antibodies to endothelial antigens: key determinants of vascular immunotargeting. *Am. J. Physiol.-Lung Cell. Mol. Physiol.* 280(6) (2001) L1335–L1347.
8. V.R. Muzykantov, Delivery of antioxidant enzyme proteins to the lung. *Antioxid. Redox Signal.* 3(1) (2001) 39–62.
9. H.L. Kutscher, P. Chao, M. Deshmukh, Y. Singh, P. Hu, L.B. Joseph, D.C. Reimer, S. Stein, D.L. Laskin, P.J. Sinko, Threshold size for optimal passive pulmonary targeting and retention of rigid microparticles in rats. *J. Controlled Release.* 143(1) (2010) 31–37.
10. S. Gebb, T. Stevens, On lung endothelial cell heterogeneity. *Microvasc. Res.* 68(1) (2004) 1–12.
11. C.D. Ochoa, S.W. Wu, T. Stevens, New developments in lung endothelial heterogeneity: von Willebrand factor, P-selectin, and the Weibel-Palade body. *Semin. Thromb. Hemost.* 36(3) (2010) 301–308.
12. A.C. Gasic, G. McGuire, S. Krater, A.I. Farhood, M.A. Goldstein, C.W. Smith, M.L. Entman, A.A. Taylor, Hydrogen peroxide pretreatment of perfused canine vessels induces ICAM-1 and CD18-dependent neutrophil adherence. *Circulation.* 84(5) (1991) 2154–2166.
13. R. Rothlein, C. Wegner, Role of intercellular adhesion molecule-1 in the inflammatory response. *Kidney Int.* 41(3) (1992) 617–619.
14. S.M. Albelda, Endothelial and epithelial cell adhesion molecules. *Am. J. Respir. Cell. Mol. Biol.* 4(3) (1991) 195–203.
15. M.T. Nakada, K. Amin, M. Christofidou-Solomidou, C.D. O'Brien, J. Sun, I. Gurubhagavatula, G.A. Heavner, A.H. Taylor, C. Paddock, Q.H. Sun, J.L. Zehnder, P.J. Newman, S.M. Albelda, H.M. DeLisser, Antibodies against the first Ig-like domain of human platelet endothelial cell adhesion molecule-1 (PECAM-1) that inhibit PECAM-1-dependent homophilic adhesion block in vivo neutrophil recruitment. *J. Immunol.* 164(1) (2000) 452–462.
16. A. Scherpereel, R. Wiewrodt, M. Christofidou-Solomidou, R. Gervais, J.C. Murciano, S.M. Albelda, V.R. Muzykantov, Cell-selective intracellular delivery of a foreign enzyme to endothelium in vivo using vascular immunotargeting. *FASEB J.* 15(2) (2001) 416–426.
17. M. Christofidou-Solomidou, A. Scherpereel, R. Wiewrodt, K. Ng, T. Sweitzer, E. Arguiri, V. Shuvaev, C.C. Solomides, S.M. Albelda, V.R. Muzykantov, PECAM-directed delivery of catalase to endothelium protects against pulmonary vascular oxidative stress. *Am. J. Physiol.-Lung Cell. Mol. Physiol.* 285(2) (2003) L283–L292.

18. S. Li, Y.D. Tan, E. Viroonchatapan, B.R. Pitt, L. Huang, Targeted gene delivery to pulmonary endothelium by anti-PECAM antibody. *Am. J. Physiol.-Lung Cell. Mol. Physiol.* 278(3) (2000) L504–L511.
19. B.D. Kozower, M. Christofidou-Solomidou, T.D. Sweitzer, S. Muro, D.G. Buerk, C.C. Solomides, S.M. Albelda, G.A. Patterson, V.R. Muzykantov, Immunotargeting of catalase to the pulmonary endothelium alleviates oxidative stress and reduces acute lung transplantation injury. *Nat. Biotechnol.* 21(4) (2003) 392–398.
20. T.K. Kishimoto, R. Rothlein, Integrins, ICAMs, and selectins: role and regulation of adhesion molecules in neutrophil recruitment to inflammatory sites. *Adv. Pharmacol.* 25 (1994) 117–169.
21. M.R. Sarantos, S. Raychaudhuri, A.F.H. Lum, D.E. Staunton, S.I. Simon, Leukocyte function-associated antigen 1-mediated adhesion stability is dynamically regulated through affinity and valency during bond formation with intercellular adhesion molecule-1. *J. Biol. Chem.* 280(31) (2005) 28290–28298.
22. G.A. Zimmerman, T.M. Mclntyre, S.M. Prescott, Adhesion and signaling in vascular cell-cell interactions. *J. Clin. Invest.* 98(8) (1996) 1699–1702.
23. J.C. Murciano, S. Muro, L. Koniaris, M. Christofidou-Solomiclou, D.W. Harshaw, S.M. Albelda, D.N. Granger, D.B. Cines, V.R. Muzykantov, ICAM-directed vascular immunotargeting of antithrombotic agents to the endothelial luminal surface. *Blood.* 101(10) (2003) 3977–3984.
24. E.N. Atochina, I.V. Balyasnikova, S.M. Danilov, D.N. Granger, A.B. Fisher, V.R. Muzykantov, Immunotargeting of catalase to ACE or ICAM-1 protects perfused rat lungs against oxidative stress. *Am. J. Physiol.* 275(4) (1998) L806–L817.
25. S. Muro, X. Cui, C. Gajewski, J.C. Murciano, V.R. Muzykantov, M. Koval, Slow intracellular trafficking of catalase nanoparticles targeted to ICAM-1 protects endothelial cells from oxidative stress. *Am. J. Physiol. Cell. Physiol.* 285(5) (2003) C1339–C1347.
26. C. Garnacho, R. Dhami, E. Simone, T. Dziubla, J. Leferovich, E.H. Schuchman, V. Muzykantov, S. Muro, Delivery of acid sphingomyelinase in normal and niemann-pick disease mice using intercellular adhesion molecule-1-targeted polymer nanocarriers. *J. Pharmacol. Exp. Ther.* 325(2) (2008) 400–408.
27. D.E. Sasso, M.A. Gionfriddo, R.S. Thrall, S.I. Syrbu, H.M. Smilowitz, R.E. Weiner, Biodistribution of indium-111-labeled antibody directed against intercellular adhesion molecule-1. *J. Nucl. Med.* 37(4) (1996) 656–661.

28. R. Rossin, S. Muro, M.J. Welch, V.R. Muzykantov, D.P. Schuster, In vivo imaging of Cu-64-labeled polymer nanoparticles targeted to the lung endothelium. *J. Nucl. Med.* 49(1) (2008) 103–111.

29. F.S. Villanueva, R.J. Jankowski, S. Klibanov, M.L. Pina, S.M. Alber, S.C. Watkins, G.H. Brandenburger, W.R. Wagner, Microbubbles targeted to intercellular adhesion molecule-1 bind to activated coronary artery endothelial cells. *Circulation.* 98(1) (1998) 1–5.

30. H. Heitsch, S. Brovkovych, T. Malinski, G. Wiemer, Angiotensin-(1–7)-stimulated nitric oxide and superoxide release from endothelial cells. *Hypertension.* 37(1) (2001) 72–76.

31. V.R. Muzykantov, E.N. Atochina, H. Ischiropoulos, S.M. Danilov, A.B. Fisher, Immunotargeting of antioxidant enzyme to the pulmonary endothelium. *Proc. Natl. Acad. Sci. U. S. A.* 93(11) (1996) 5213–5218.

32. K. Nowak, S. Weih, R. Metzger, R.F. Albrecht, S. Post, P. Hohenberger, M.M. Gebhard, S.M. Danilov, Immunotargeting of catalase to lung endothelium via anti-angiotensin-converting enzyme antibodies attenuates ischemia-reperfusion injury of the lung in vivo. *Am. J. Physiol.-Lung Cell. Mol. Physiol.* 293(1) (2007) L162–L169.

33. P.N. Reynolds, K.R. Zinn, V.D. Gavrilyuk, I.V. Balyasnikova, B.E. Rogers, D.J. Buchsbaum, M.H. Wang, D.J. Miletich, W.E. Grizzle, J.T. Douglas, S.M. Danilov, D.T. Curiel, A targetable, injectable adenoviral vector for selective gene delivery to pulmonary endothelium in vivo. *Mol. Ther.* 2(6) (2000) 562–578.

34. C.T. Esmon, The roles of protein C and thrombomodulin in the regulation of blood coagulation. *J. Biol. Chem.* 264(9) (1989) 4743–4746.

35. M. Christofidou-Solomidou, S. Kennel, A. Scherpereel, R. Wiewrodt, C.C. Solomides, G.G. Pietra, J.C. Murciano, S.A. Shah, H. Ischiropoulos, S.M. Albeda, V.R. Muzykantov, Vascular immunotargeting of glucose oxidase to the endothelial antigens induces distinct forms of oxidant acute lung injury - Targeting to thrombomodulin, but not to PECAM-1, causes pulmonary thrombosis and neutrophil transmigration. *Am. J. Pathol.* 160(3) (2002) 1155–1169.

36. V.S. Trubetskoy, V.P. Torchilin, S.J. Kennel, L. Huang, Use of N-terminal modified poly(L-lysine)-antibody conjugate as a carrier for targeted gene delivery in mouse lung endothelial cells. Bioconjugate Chem. 3(4) (1992) 323–327.

37. T.A. Springer, Adhesion receptors of the immune system. *Nature.* 346(6283) (1990) 425–434.

38. R.J. Kok, M. Everts, S.A. Asgeirsdottir, D.K.F. Meijer, G. Molema, Cellular handling of a dexamethasone-anti-E-selectin immunoconjugate by

activated endothelial cells: comparison with free dexamethasone. *Pharm. Res.* 19(11) (2002) 1730–1735.

39. S. Kessner, A. Krause, U. Rothe, G. Bendas, Investigation of the cellular uptake of E-Selectin-targeted immunoliposomes by activated human endothelial cells. *Biochim. Biophys. Acta-Biomembr.* 1514(2) (2001) 177–190.
40. O.A. Harari, T.J. Wickham, C.J. Stocker, I. Kovesdi, D.M. Segal, T.Y. Huehns, C. Sarraf, D.O. Haskard, Targeting an adenoviral gene vector to cytokine-activated vascular endothelium via E-selectin. *Gene Ther.* 6(5) (1999) 801–807.
41. J.R. Lindner, J. Song, J. Christiansen, A.L. Klibanov, F. Xu, K. Ley, Ultrasound assessment of inflammation and renal tissue injury with microbubbles targeted to P-selectin. *Circulation.* 104(17) (2001) 2107–2112.
42. E.T.M. Keelan, A.A. Harrison, P.T. Chapman, R.M. Binns, A.M. Peters, D.O. Haskard, Imaging vascular endothelial activation: an approach using radiolabeled monoclonal antibodies against the endothelial cell adhesion molecule E-selectin. *J. Nucl. Med.* 35(2) (1994) 276–281.
43. K.A. Kelly, J.R. Allport, A. Tsourkas, V.R. Shinde-Patil, L. Josephson, R. Weissleder, Detection of vascular adhesion molecule-1 expression using a novel multimodal nanoparticle. *Circ. Res.* 96(3) (2005) 327–336.
44. P. Holig, M. Bach, T. Volkel, T. Nahde, S. Hoffmann, R. Muller, R.E. Kontermann, Novel RGD lipopeptides for the targeting of liposomes to integrin-expressing endothelial and melanoma cells. *Protein Eng. Des. Sel.* 17(5) (2004) 433–441.
45. Y. Wang, X. Wang, Y. Zhang, S. Yang, J. Wang, X. Zhang, Q. Zhang, RGD-modified polymeric micelles as potential carriers for targeted delivery to integrin-overexpressing tumor vasculature and tumor cells. *J. Drug Target.* 17(6) (2009) 459–467.
46. A.J. Schraaa, R.J. Kok, A.D. Berendsen, H.E. Moorlag, E.J. Bos, D.K.F. Meijer, L. de Leij, G. Molema, Endothelial cells internalize and degrade RGD-modified proteins developed for tumor vasculature targeting. *J. Controlled Release.* 83(2) (2002) 241–251.
47. X.Y. Chen, E. Sievers, Y.P. Hou, R. Park, M. Tohme, R. Bart, R. Bremner, J.R. Bading, P.S. Conti, Integrin alpha(V)beta(3)-targeted imaging of lung cancer. *Neoplasia.* 7(3) (2005) 271–279.
48. A.P.C.A. Janssen, R.M. Schiffelers, T.L.M. ten Hagen, G.A. Koning, A.J. Schraa, R.J. Kok, G. Storm, G. Molema, Peptide-targeted PEG-liposomes in anti-angiogenic therapy. *Int. J. Pharm.* 254(1) (2003) 55–58.

49. L. Zhang, M.J. Crossley, N.E. Dixon, P.J. Ellis, M.L. Fisher, G.F. King, P.E. Lilley, D. MacLachlan, R.J. Pace, H.C. Freeman, Spectroscopic identification of a dinuclear metal centre in manganese(II)-activated aminopeptidase P from Escherichia coli: implications for human prolidase. *J. Biol. Inorg. Chem.* 3(5) (1998) 470–483.

50. P. Oh, Y. Li, J.Y. Yu, E. Durr, K.M. Krasinska, L.A. Carver, J.E. Testa, J.E. Schnitzer, Subtractive proteomic mapping of the endothelial surface in lung and solid tumours for tissue-specific therapy. *Nature.* 429(6992) (2004) 629–635.

51. S.J. White, S.A. Nicklin, T. Sawamura, A.H. Baker, Identification of peptides that target the endothelial cell-specific LOX-1 receptor. *Hypertension.* 37(2) (2001) 449–455.

52. S. Ishino, T. Mukai, Y. Kuge, N. Kume, M. Ogawa, N. Takai, J. Kamihashi, M. Shiomi, M. Minami, T. Kita, H. Saji, Targeting of lectinlike oxidized low-density lipoprotein receptor 1 (LOX-1) with 99mTc-labeled anti-LOX-1 antibody: potential agent for imaging of vulnerable plaque. *J. Nucl. Med.* 49(10) (2008) 1677–1685.

53. K.A. Massey, J.E. Schnitzer, Targeting and imaging signature caveolar molecules in lungs. *Proc. Am. Thorac. Soc.* 6(5) (2009) 419–430.

54. L.H.K. Lim, S. Pervaiz, Annexin 1: the new face of an old molecule. *FASEB J.* 21(4) (2007) 968–975.

55. D. Neri, R. Bicknell, Tumour vascular targeting. *Nat. Rev. Cancer.* 5(6) (2005) 436–446.

56. T.A. John, S.M. Vogel, C. Tiruppathi, A.B. Malik, R.D. Minshall, Quantitative analysis of albumin uptake and transport in the rat microvessel endothelial monolayer. *Am. J. Physiol. Lung. Cell. Mol. Physiol.* 284(1) (2003) L187–196.

57. E. Ruoslahti, M. Pierschbacher, New perspectives in cell adhesion: RGD and integrins. *Science.* 238(4826) (1987) 491–497.

58. Y.F. Zhang, J.C. Wang, D.Y. Bian, X. Zhang, Q. Zhang, Targeted delivery of RGD-modified liposomes encapsulating both combretastatin A-4 and doxorubicin for tumor therapy: *in vitro* and *in vivo* studies. *Eur. J. Pharm. Biopharm.* 74(3) (2010) 467–473.

59. S.Z. Li, M.J. McGuire, M. Lin, Y.H. Liu, T. Oyama, X.K. Sun, K.C. Brown, Synthesis and characterization of a high-affinity alpha(v)beta(6)-specific ligand for in vitro and in vivo applications. *Mol. Cancer Ther.* 8(5) (2009) 1239–1249.

60. E. Koivunen, D.A. Gay, E. Ruoslahti, Selection of peptides binding to the alpha 5 beta 1 integrin from phage display library. *J. Biol. Chem.* 268(27) (1993) 20205–20210.

61. N. Zhang, C. Chittasupho, C. Duangrat, T.J. Siahaan, C. Berkland, PLGA nanoparticle–peptide conjugate effectively targets intercellular cell-adhesion molecule-1. *Bioconjug. Chem.* 19(1) (2008) 145–152.

62. J.O. McNamara, E.R. Andrechek, Y. Wang, K.D. Viles, R.E. Rempel, E. Gilboa, B.A. Sullenger, P.H. Giangrande, Cell type-specific delivery of siRNAs with aptamer-siRNA chimeras. *Nat. Biotech.* 24(8) (2006) 1005–1015.

63. N.S. Que-Gewirth, B.A. Sullenger, Gene therapy progress and prospects: RNA aptamers. *Gene. Ther.* 14(4) (2007) 283–291.

64. K. Maruyama, E. Holmberg, S.J. Kennel, A. Klibanov, V.P. Torchilin, L. Huang, Characterization of in vivo immunoliposome targeting to pulmonary endothelium. *J. Pharm. Sci.* 79(11) (1990) 978–984.

65. J.H. Finigan, A. Boueiz, E. Wilkinson, R. Damico, J. Skirball, H.H. Pae, M. Damarla, E. Hasan, D.B. Pearse, S.P. Reddy, D.N. Grigoryev, C. Cheadle, C.T. Esmon, J.G. Garcia, P.M. Hassoun, Activated protein C protects against ventilator-induced pulmonary capillary leak. *Am. J. Physiol. Lung. Cell. Mol. Physiol.* 296(6) (2009) L1002–L1011.

66. J.S. Beckman, R.L. Minor, C.W. White, J.E. Repine, G.M. Rosen, B.A. Freeman, Superoxide dismutase and catalase conjugated to polyethylene glycol increases endothelial enzyme activity and oxidant resistance. *J. Biol. Chem.* 263(14) (1988) 6884–6892.

67. F. Herve, N. Ghinea, J.M. Scherrmann, CNS delivery via adsorptive transcytosis. *AAPS J.* 10(3) (2008) 455–472.

68. H. Xia, Q. Mao, B.L. Davidson, The HIV tat protein transduction domain improves the biodistribution of bold beta-glucuronidase expressed from recombinant viral vectors. *Nat. Biotech.* 19(7) (2001) 640–644.

69. V.P. Torchilin, R. Rammohan, V. Weissig, T.S. Levchenko, TAT peptide on the surface of liposomes affords their efficient intracellular delivery even at low temperature and in the presence of metabolic inhibitors. *Proc. Natl. Acad. Sci. U. S. A.* 98(15) (2001) 8786–8791.

70. S. Console, C. Marty, C. Garcia-Echeverria, R. Schwendener, K. Ballmer-Hofer, Antennapedia and HIV transactivator of transcription (TAT) "protein transduction domains" promote endocytosis of high molecular weight cargo upon binding to cell surface glycosaminoglycans. *J. Biol. Chem.* 278(37) (2003) 35109–35114.

71. J.M. de la Fuente, C.C. Berry, Tat peptide as an efficient molecule to translocate gold nanoparticles into the cell nucleus. *Bioconjugate Chem.* 16(5) (2005) 1176–1180.

72. I.D. Alves, C.Y. Jiao, S. Aubry, B. Aussedat, F. Burlina, G. Chassaing, S. Sagan, Cell biology meets biophysics to unveil the different mechanisms

of penetratin internalization in cells. *Biochim. Biophys. Acta.* (2010) doi:10.1016/j.bbamem.2010.02.009.

73. M.S. Zhou, I. Hernandez Schulman, P.J. Pagano, E.A. Jaimes, L. Raij, Reduced NAD(P)H oxidase in low renin hypertension: link among angiotensin II, atherogenesis, and blood pressure. *Hypertension.* 47(1) (2006) 81–86.
74. S.D. Conner, S.L. Schmid, Regulated portals of entry into the cell. *Nature.* 422(6927) (2003) 37–44.
75. C. Lamaze, S.L. Schmid, The emergence of clathrin-independent pinocytic pathways. *Curr. Opin. Cell. Biol.* 7(4) (1995) 573–580.
76. I.A. Khalil, K. Kogure, H. Akita, H. Harashima, Uptake pathways and subsequent intracellular trafficking in nonviral gene delivery. *Pharmacol. Rev.* 58(1) (2006) 32–45.
77. S.L. Schmid, Clathrin-coated vesicle formation and protein sorting: an integrated process. *Annu. Rev. Biochem.* 66 (1997) 511–548.
78. Y. Liu, S.C. Steiniger, Y. Kim, G.F. Kaufmann, B. Felding-Habermann, K.D. Janda, Mechanistic studies of a peptidic GRP78 ligand for cancer cell-specific drug delivery. *Mol. Pharm.* 4(3) (2007) 435–447.
79. W. Ke, K. Shao, R. Huang, L. Han, Y. Liu, J. Li, Y. Kuang, L. Ye, J. Lou, C. Jiang, Gene delivery targeted to the brain using an angiopep-conjugated polyethyleneglycol-modified polyamidoamine dendrimer. *Biomaterials.* 30(36) (2009) 6976–6985.
80. S. Sohrab, D.N. Petrusca, A.D. Lockett, K.S. Schweitzer, N.I. Rush, Y. Gu, K. Kamocki, J. Garrison, I. Petrache, Mechanism of alpha-1 antitrypsin endocytosis by lung endothelium. *FASEB J.* 23(9) (2009) 3149–3158.
81. D. Predescu, G.E. Palade, Plasmalemmal vesicles represent the large pore system of continuous microvascular endothelium. *Am. J. Physiol.* 265(2) (1993) H725–H733.
82. M. Murata, J. Peranen, R. Schreiner, F. Wieland, T.V. Kurzchalia, K. Simons, VIP21/caveolin is a cholesterol-binding protein. *Proc. Natl. Acad. Sci. U. S. A.* 92(22) (1995) 10339–10343.
83. Z. Wang, C. Tiruppathi, R.D. Minshall, A.B. Malik, Size and dynamics of caveolae studied using nanoparticles in living endothelial cells. *ACS Nano.* 3(12) (2009) 4110–4116.
84. K.A. Massey, J.E. Schnitzer, Targeting and imaging signature caveolar molecules in lungs. *Proc. Am. Thorac. Soc.* 6(5) (2009) 419–430.
85. P. Wang, Y. Xue, X. Shang, Y. Liu, Diphtheria toxin mutant CRM197-mediated transcytosis across blood-brain barrier in vitro. *Cell. Mol. Neurobiol.* 30(5) (2010) 717–725.

86. R.V. Stan, Endocytosis pathways in endothelium: how many? *Am. J. Physiol. Lung Cell. Mol. Physiol.* 290(5) (2006) L806–L808.

87. S. Muro, R. Wiewrodt, A. Thomas, L. Koniaris, S.M. Albelda, V.R. Muzykantov, M. Koval, A novel endocytic pathway induced by clustering endothelial ICAM-1 or PECAM-1. *J. Cell. Sci.* 116(8) (2003) 1599–1609.

88. S. Muro, E.H. Schuchman, V.R. Muzykantov, Lysosomal enzyme delivery by ICAM-1-targeted nanocarriers bypassing glycosylation- and clathrin-dependent endocytosis. *Mol. Ther.* 13(1) (2006) 135–141.

89. D.T. Curiel, S. Agarwal, E. Wagner, M. Cotten, Adenovirus enhancement of transferrin-polylysine-mediated gene delivery. *Proc. Natl. Acad. Sci. U. S. A.* 88(19) (1991) 8850–8854.

90. H. Mizuguchi, T. Nakagawa, M. Nakanishi, S. Imazu, S. Nakagawa, T. Mayumi, Efficient gene transfer into mammalian cells using fusogenic liposome. *Biochem. Biophys. Res. Commun.* 218(1) (1996) 402–407.

91. W. Li, F. Nicol, F.C. Szoka, GALA: a designed synthetic pH-responsive amphipathic peptide with applications in drug and gene delivery. *Adv. Drug Deliv. Rev.* 56(7) (2004) 967–985.

92. O. Zelphati, F.C. Szoka, Mechanism of oligonucleotide release from cationic liposomes. *Proc. Natl. Acad. Sci. U. S. A.* 93(21) (1996) 11493–11498.

93. J. Connor, L. Huang, Efficient cytoplasmic delivery of a fluorescent dye by pH-sensitive immunoliposomes. *J. Cell Biol.* 101(2) (1985) 582–589.

94. N. Murthy, J.R. Robichaud, D.A. Tirrell, P.S. Stayton, A.S. Hoffman, The design and synthesis of polymers for eukaryotic membrane disruption. *J. Controlled Release.* 61(1–2) (1999) 137–143.

95. N.D. Sonawane, F.C. Szoka, Jr., A.S. Verkman, Chloride accumulation and swelling in endosomes enhances DNA transfer by polyamine-DNA polyplexes. *J. Biol. Chem.* 278(45) (2003) 44826–44831.

96. Y. Hu, T. Litwin, A.R. Nagaraja, B. Kwong, J. Katz, N. Watson, D.J. Irvine, Cytosolic delivery of membrane-impermeable molecules in dendritic cells using pH-responsive core-shell nanoparticles. *Nano Letters.* 7(10) (2007) 3056–3064.

97. I.G. Macara, Transport into and out of the nucleus. *Microbiol. Mol. Biol. Rev.* 65(4) (2001) 570–594.

98. F.M. Munkonge, D.A. Dean, E. Hillery, U. Griesenbach, E.W. Alton, Emerging significance of plasmid DNA nuclear import in gene therapy. *Adv. Drug Deliv. Rev.* 55(6) (2003) 749–760.

99. A.M. Miller, D.A. Dean, Cell-specific nuclear import of plasmid DNA in smooth muscle requires tissue-specific transcription factors and DNA sequences. *Gene Ther.* 15(15) (2008) 1107–1115.

100. A.M. Miller, F.M. Munkonge, E.W. Alton, D.A. Dean, Identification of protein cofactors necessary for sequence-specific plasmid DNA nuclear import. *Mol. Ther.* 17(11) (2009) 1897–1903.

101. D. Mudhakir, H. Harashima, Learning from the viral journey: how to enter cells and how to overcome intracellular barriers to reach the nucleus. *AAPS J.* 11(1) (2009) 65–77.

102. F. Qing, T.J. McCarthy, J. Markham, D.P. Schuster, Pulmonary ngiotensin-converting enzyme (ACE) binding and inhibition in humans — a positron emission tomography study. *Am. J. Respir. Crit. Care Med.* 161(6) (2000) 2019–2025.

103. F.J. Femia, K.P. Maresca, S.M. Hillier, C.N. Zimmerman, J.L. Joyal, J.A. Barrett, O. Aras, V. Dilsizian, W.C. Eckelman, J.W. Babich, Synthesis and evaluation of a series of Tc-99m(CO)(3)(+) lisinopril complexes for in vivo imaging of angiotensin-converting enzyme expression. *J. Nucl. Med.* 49(6) (2008) 970–977.

104. B.S. Ding, T. Dziubla, V.V. Shuvaev, S. Muro, V.R. Muzykantov, Advanced drug delivery systems that target the vascular endothelium. *Mol. Interv.* 6(2) (2006) 98–112.

105. S.S. Wong, *Chemistry of Protein Conjugation and Cross-Linking*, CRC Press, Boca Raton, 1991.

106. V.V. Shuvaev, M. Christofidou-Solomidou, A. Scherpereel, E. Simone, E. Arguiri, S. Tliba, J. Pick, S. Kennel, S.M. Albelda, V.R. Muzykantov, Factors modulating the delivery and effect of enzymatic cargo conjugated with antibodies targeted to the pulmonary endothelium. *J. Controlled Release.* 118(2) (2007) 235–244.

107. K. Nowak, C. Hanusch, K. Nicksch, R.P. Metzger, G. Beck, M.M. Gebhard, P. Hohenberger, S.M. Danilov, Pre-ischaemic conditioning of the pulmonary endothelium by immunotargeting of catalase via angiotensin-converting-enzyme antibodies. *Eur. J. Cardiothorac. Surg.* 37(4) (2010) 859–863.

108. B.S. Ding, C. Gottstein, A. Grunow, A. Kuo, K. Ganguly, S.M. Albelda, D.B. Cines, V.R. Muzykantov, Endothelial targeting of a recombinant construct fusing a PECAM-1 single-chain variable antibody fragment (scFv) with prourokinase facilitates prophylactic thrombolysis in the pulmonary vasculature. *Blood.* 106(13) (2005) 4191–4198.

109. B.S. Ding, N. Hong, J.C. Murciano, K. Ganguly, C. Gottstein, M. Christofidou-Solomidou, S.M. Albelda, A.B. Fisher, D.B. Cines, V.R.

Muzykantov, Prophylactic thrombolysis by thrombin-activated latent prourokinase targeted to PECAM-1 in the pulmonary vasculature. *Blood.* 111(4) (2008) 1999–2006.

110. V.R. Muzykantov, M. Christofidou-Solomidou, I. Balyasnikova, D.W. Harshaw, L. Schultz, A.B. Fisher, S.M. Albelda, Streptavidin facilitates internalization and pulmonary targeting of an anti-endothelial cell antibody (platelet-endothelial cell adhesion molecule 1): A strategy for vascular immunotargeting of drugs. *Proc. Natl. Acad. Sci. U. S. A.* 96(5) (1999) 2379–2384.

111. E.A. Simone, T.D. Dziubla, F. Colon-Gonzalez, D.E. Discher, V.R. Muzykantov, Effect of polymer amphiphilicity on loading of a therapeutic enzyme into protective filamentous and spherical polymer nanocarriers. *Biomacromolecules.* 8(12) (2007) 3914–3921.

112. T.D. Dziubla, A. Karim, V.R. Muzykantov, Polymer nanocarriers protecting active enzyme cargo against proteolysis. *J. Controlled Release.* 102(2) (2005) 427–439.

113. E.A. Simone, T.D. Dziubla, E. Arguiri, V. Vardon, V.V. Shuvaev, M. Christofidou-Solomidou, V.R. Muzykantov, Loading PEG-catalase into filamentous and spherical polymer nanocarriers. *Pharm. Res.* 26(1) (2009) 250–260.

114. S. Muro, T. Dziubla, W. Qiu, J. Leferovich, X. Cui, E. Berk, V.R. Muzykantov, Endothelial targeting of high-affinity multivalent polymer nanocarriers directed to intercellular adhesion molecule 1. *J. Pharmacol. Exp. Ther.* 317(3) (2006) 1161–1169.

115. S. Muro, C. Gajewski, M. Koval, V.R. Muzykantov, ICAM-1 recycling in endothelial cells: a novel pathway for sustained intracellular delivery and prolonged effects of drugs. *Blood.* 105(2) (2005) 650–658.

116. S. Muro, C. Garnacho, J.A. Champion, J. Leferovich, C. Gajewski, E.H. Schuchman, S. Mitragotri, V.R. Muzykantov, Control of endothelial targeting and intracellular delivery of therapeutic enzymes by modulating the size and shape of ICAM-1-targeted carriers. *Mol. Ther.* 16(8) (2008) 1450–1458.

117. E.A. Simone, T.D. Dziubla, D.E. Discher, V.R. Muzykantov, Filamentous polymer nanocarriers of tunable stiffness that encapsulate the therapeutic enzyme catalase. *Biomacromolecules.* 10(6) (2009) 1324–1330.

118. K.L. Brigham, B. Meyrick, B. Christman, M. Magnuson, G. King, L.C. Berry, Jr., *In vivo* transfection of murine lungs with a functioning prokaryotic gene using a liposome vehicle. *Am. J. Med. Sci.* 298(4) (1989) 278–281.

119. P.L. Felgner, T.R. Gadek, M. Holm, R. Roman, H.W. Chan, M. Wenz, J.P. Northrop, G.M. Ringold, M. Danielsen, Lipofection: a highly efficient, lipid-mediated DNA-transfection procedure. *Proc. Natl. Acad. Sci. U. S. A.* 84(21) (1987) 7413–7417.
120. T.A. Hazinski, P.A. Ladd, C.A. DeMatteo, Localization and induced expression of fusion genes in the rat lung. *Am. J. Respir. Cell. Mol. Biol.* 4(3) (1991) 206–209.
121. K.L. Brigham, B. Meyrick, B. Christman, J.T. Conary, G. King, L.C. Berry, Jr., M.A. Magnuson, Expression of human growth hormone fusion genes in cultured lung endothelial cells and in the lungs of mice. *Am. J. Respir. Cell. Mol. Biol.* 8(2) (1993) 209–213.
122. J.T. Conary, R.E. Parker, B.W. Christman, R.D. Faulks, G.A. King, B.O. Meyrick, K.L. Brigham, Protection of rabbit lungs from endotoxin injury by *in vivo* hyperexpression of the prostaglandin G/H synthase gene. *J. Clin. Invest.* 93(4) (1994) 1834–1840.
123. N. Zhu, D. Liggitt, Y. Liu, R. Debs, Systemic gene expression after intravenous DNA delivery into adult mice. *Science.* 261(5118) (1993) 209–211.
124. S. Li, L. Huang, *In vivo* gene transfer via intravenous administration of cationic lipid-protamine-DNA (LPD) complexes. *Gene Ther.* 4(9) (1997) 891–900.
125. Z. Ma, J.L. Zhang, S. Alber, J. Dileo, Y. Negishi, D. Stolz, S. Watkins, L. Huang, B. Pitt, S. Li, Lipid-mediated delivery of oligonucleotide to pulmonary endothelium. *Am. J. Respir. Cell Mol. Biol.* 27(2) (2002) 151–159.
126. Y. Liu, L.C. Mounkes, H.D. Liggitt, C.S. Brown, I. Solodin, T.D. Heath, R.J. Debs, Factors influencing the efficiency of cationic liposome-mediated intravenous gene delivery. *Nat. Biotechnol.* 15(2) (1997) 167–173.
127. A.E. Canonico, J.T. Conary, B.O. Meyrick, K.L. Brigham, Aerosol and intravenous transfection of human alpha 1-antitrypsin gene to lungs of rabbits. *Am. J. Respir. Cell. Mol. Biol.* 10(1) (1994) 24–29.
128. D.W. Muller, D. Gordon, H. San, Z. Yang, V.J. Pompili, G.J. Nabel, E.G. Nabel, Catheter-mediated pulmonary vascular gene transfer and expression. *Circ. Res.* 75(6) (1994) 1039–1049.
129. F. Liu, H. Qi, L. Huang, D. Liu, Factors controlling the efficiency of cationic lipid-mediated transfection in vivo via intravenous administration. *Gene Ther.* 4(6) (1997) 517–523.
130. N.S. Templeton, D.D. Lasic, P.M. Frederik, H.H. Strey, D.D. Roberts, G.N. Pavlakis, Improved DNA: liposome complexes for increased systemic delivery and gene expression. *Nat. Biotechnol.* 15(7) (1997) 647–652.

131. Y.K. Song, F. Liu, D. Liu, Enhanced gene expression in mouse lung by prolonging the retention time of intravenously injected plasmid DNA. *Gene Ther.* 5(11) (1998) 1531–1537.

132. S. Kawakami, Y. Ito, S. Fumoto, F. Yamashita, M. Hashida, Enhanced gene expression in lung by a stabilized lipoplex using sodium chloride for complex formation. *J. Gene. Med.* 7(12) (2005) 1526–1533.

133. N.Y. Shi, W.M. Pardridge, Noninvasive gene targeting to the brain. *Proc. Natl. Acad. Sci. U. S. A.* 97(13) (2000) 7567–7572.

134. L. Xu, P. Frederik, K.F. Pirollo, W.H. Tang, A. Rait, L.M. Xiang, W.Q. Huang, I. Cruz, Y.Z. Yin, E.H. Chang, Self-assembly of a virus-mimicking nanostructure system for efficient tumor-targeted gene delivery. *Hum. Gene Ther.* 13(3) (2002) 469–481.

135. S.C. Semple, S.K. Klimuk, T.O. Harasym, N. Dos Santos, S.M. Ansell, K.F. Wong, N. Maurer, H. Stark, P.R. Cullis, M.J. Hope, P. Scherrer, Efficient encapsulation of antisense oligonucleotides in lipid vesicles using ionizable aminolipids: formation of novel small multilamellar vesicle structures. *Biochim. Biophys. Acta-Biomembr.* 1510(1–2) (2001) 152–166.

136. A. Wilson, W. Zhou, H.C. Champion, S. Alber, Z.L. Tang, S. Kennel, S. Watkins, L. Huang, B. Pitt, S. Li, Targeted delivery of oligodeoxynucleotides to mouse lung endothelial cells *in vitro* and *in vivo*. *Mol. Ther.* 12(3) (2005) 510–518.

137. N.S. Yew, K.X. Wang, M. Przybylska, R.G. Bagley, M. Stedman, J. Marshall, R.K. Scheule, S.H. Cheng, Contribution of plasmid DNA to inflammation in the lung after administration of cationic lipid : pDNA complexes. *Hum. Gene Ther.* 10(2) (1999) 223–234.

138. S. Li, S.-P. Wu, M. Whitmore, E.J. Loeffert, L. Wang, S.C. Watkins, B.R. Pitt, L. Huang, Effect of immune response on gene transfer to the lung via systemic administration of cationic lipidic vectors. *Am. J. Physiol. Lung Cell. Mol. Physiol.* 276(5) (1999) L796–L804.

139. Y.D. Tan, S. Li, B.R. Pitt, L. Huang, The inhibitory role of CpG immunostimulatory motifs in cationic lipid vector-mediated transgene expression in vivo. *Hum. Gene Ther.* 10(13) (1999) 2153–2161.

140. C.R. Hofman, J.P. Dileo, Z. Li, S. Li, L. Huang, Efficient in vivo gene transfer by PCR amplified fragment with reduced inflammatory activity. *Gene Ther.* 8(1) (2001) 71–74.

141. N.S. Yew, H.M. Zhao, I.H. Wu, A. Song, J.D. Tousignant, M. Przybylska, S.H. Cheng, Reduced inflammatory response to plasmid DNA vectors by elimination and inhibition of immunostimulatory CpG motifs. *Mol. Ther.* 1(3) (2000) 255–262.

142. Y.D. Tan, F. Liu, Z.Y. Li, S. Li, L. Huang, Sequential injection of cationic liposome and plasmid DNA effectively transfects the lung with minimal inflammatory toxicity. *Mol. Ther.* 3(5) (2001) 673–682.

143. A.G. Ziady, T. Ferkol, D.V. Dawson, D.H. Perlmutter, P.B. Davis, Chain length of the polylysine in receptor-targeted gene transfer complexes affects duration of reporter gene expression both *in vitro* and *in vivo*. *J. Biol. Chem.* 274(8) (1999) 4908–4916.

144. A.G. Ziady, C.R. Gedeon, T. Miller, W. Quan, J.M. Payne, S.L. Hyatt, T.L. Fink, O. Muhammad, S. Oette, T. Kowalczyk, M.K. Pasumarthy, R.C. Moen, M.J. Cooper, P.B. Davis, Transfection of airway epithelium by stable PEGylated poly-L-lysine DNA nanoparticles *in vivo*. *Mol. Ther.* 8(6) (2003) 936–947.

145. T. Ferkol, J.C. Perales, E. Eckman, C.S. Kaetzel, R.W. Hanson, P.B. Davis, Gene transfer into the airway epithelium of animals by targeting the polymeric immunoglobulin receptor. *J. Clin. Invest.* 95(2) (1995) 493–502.

146. O. Boussif, F. Lezoualc'h, M.A. Zanta, M.D. Mergny, D. Scherman, B. Demeneix, J.P. Behr, A versatile vector for gene and oligonucleotide transfer into cells in culture and in vivo: polyethylenimine. *Proc. Natl. Acad. Sci. U. S. A.* 92(16) (1995) 7297–7301.

147. D. Goula, C. Benoist, S. Mantero, G. Merlo, G. Levi, B.A. Demeneix, Polyethylenimine-based intravenous delivery of transgenes to mouse lung. *Gene Ther.* 5(9) (1998) 1291–1295.

148. L. Wightman, R. Kircheis, V. Rossler, S. Carotta, R. Ruzicka, M. Kursa, E. Wagner, Different behavior of branched and linear polyethylenimine for gene delivery in vitro and in vivo. *J. Gene Med.* 3(4) (2001) 362–372.

149. M. Thomas, A.M. Klibanov, Enhancing polyethylenimine's delivery of plasmid DNA into mammalian cells. *Proc. Natl. Acad. Sci. U. S. A.* 99(23) (2002) 14640–14645.

150. A. Kichler, Gene transfer with modified polyethylenimines. *J. Gene. Med.* 6(S1) (2004) S3–S10.

151. R. Kircheis, T. Blessing, S. Brunner, L. Wightman, E. Wagner, Tumor targeting with surface-shielded ligand–polycation DNA complexes. *J. Controlled Release.* 72(1–3) (2001) 165–170.

152. M.A. Gosselin, W. Guo, R.J. Lee, Efficient gene transfer using reversibly cross-linked low molecular weight polyethylenimine. *Bioconjug. Chem.* 12(6) (2001) 989–994.

153. D. Wu, Y. Liu, X. Jiang, L. Chen, C. He, S.H. Goh, K.W. Leong, Evaluation of hyperbranched poly(amino ester)s of amine constitutions similar to polyethylenimine for DNA delivery. *Biomacromolecules.* 6(6) (2005) 3166–3173.

154. A. Akinc, D.G. Anderson, D.M. Lynn, R. Langer, Synthesis of poly(beta-amino ester)s optimized for highly effective gene delivery. *Bioconjug. Chem.* 14(5) (2003) 979–988.

155. D.G. Anderson, A. Akinc, N. Hossain, R. Langer, Structure/property studies of polymeric gene delivery using a library of poly(beta-amino esters). *Mol. Ther.* 11(3) (2005) 426–434.

156. X. Gao, R. Kuruba, K. Damodaran, B.W. Day, D. Liu, S. Li, Polyhydroxylalkyleneamines: a class of hydrophilic cationic polymer-based gene transfer agents. *J. Controlled Release.* 137(1) (2009) 38–45.

157. Z. Ma, Z. Mi, A. Wilson, S. Alber, P.D. Robbins, S. Watkins, B. Pitt, S. Li, Redirecting adenovirus to pulmonary endothelium by cationic liposomes. *Gene Ther.* 9(3) (2002) 176–182.

158. P.N. Reynolds, S.A. Nicklin, L. Kaliberova, B.G. Boatman, W.E. Grizzle, I.V. Balyasnikova, A.H. Baker, S.M. Danilov, D.T. Curiel, Combined transductional and transcriptional targeting improves the specificity of transgene expression in vivo. *Nat. Biotechnol.* 19(9) (2001) 838–842.

159. L.M. Work, H. Buning, E. Hunt, S.A. Nicklin, L. Denby, N. Britton, K. Leike, M. Odenthal, U. Drebber, M. Hallek, A.H. Baker, Vascular bed-targeted in vivo gene delivery using tropism-modified adeno-associated viruses. *Mol. Ther.* 13(4) (2006) 683–693.

160. W.J. Li, F.C. Szoka, Lipid-based nanoparticles for nucleic acid delivery. *Pharm. Res.* 24(3) (2007) 438–449.

161. D.B. Fenske, I. MacLachlan, P.R. Cullis, Long-circulating vectors for the systemic delivery of genes. *Curr. Opin. Mol. Ther.* 3(2) (2001) 153–158.

Chapter 8

Nanosystems for Selective Epithelial Barrier Targeting in Chronic Airway Diseases

Heather A. Parsons,[a] Rachel L. Damico,[b] and Venkataramana K. Sidhaye[b,*]

The [a]Departments of Medicine and [b]Division of Pulmonary and Critical Care Medicine, Johns Hopkins University, Baltimore, MD, USA
[*]vsidhay1@jhmi.edu

8.1 Introduction

The obstructive lung diseases — chronic obstructive pulmonary disease (COPD) and bronchial asthma — together are responsible for a significant portion of global morbidity and mortality, causing more than 3 million deaths worldwide each year and affecting more than 510 million people [1, 2]. And while more than 90% of these deaths occur in developing countries, obstructive lung diseases are a leading cause of death in the United States, fourth behind heart disease, cancer, and stroke [3]. They are predicted to become the third-leading cause of death by 2020. We, in the United States, spend more than $20 million per year on asthma and COPD, and that expense is expected to rise in the coming years.

In 1900, asthma was rare, but it has grown in epidemic proportions. Though the World Health Organization estimates that

Pulmonary Nanomedicine: Diagnostics, Imaging, and Therapeutics
Edited by Neeraj Vij

ISBN 978-981-4316-48-4 (Hardback), 978-981-4364-14-0 (eBook)
www.panstanford.com

asthma kills 5,000 Americans every year and 180,000 annually worldwide [1], the cause for the increase in asthma is not entirely understood. Studies indicate that its prevalence is highest in Western countries, particularly in the English-speaking nations. It is virtually absent in parts of rural Africa, and this fact has caused investigators to assess the environmental and lifestyle differences in these populations [4].

Although having asthmatic parents increases a child's risk, there seems to be a consensus that the increase in cases of asthma in Western countries is not predominantly due to genetic factors but due to the environment and lifestyle, though elements dictating this are not clear. Among the candidates is the tendency of children to spend more time indoors than did those in earlier generations, potentially resulting in increased exposure to household allergens, including dust mites, cats, and cockroaches; decreased exposure to other outdoor factors, such as vitamin D; or a combination of both.

According to one popular theory, the pulmonary immune systems of Western children are less mature as they are not conditioned to live with parasites, and so the children become more vulnerable to asthma and other allergic diseases such as hay fever and eczema. While all these hypotheses have epidemiologic evidence to support them, the biologic mechanisms underlying these are far from known.

COPD is a disease that has been affecting an exponentially increasing number of people but that has not seen anything nearing a proportionate increase in the available secondary or tertiary prevention options. As an indication of its impact on society, in 1990, COPD ranked twelfth as a cause of lost disability-adjusted life years (DALYs) — the sum of life-years lost to premature death and to disability — and it is estimated that COPD will be the fifth-leading cause of lost DALYs globally by 2020 [5]. The GOLD initiative was launched in 2001 to focus on the prevention and treatment of COPD and on the research necessary to improve these, but thus far, treatment options remain limited.

We currently have only a few therapies available for the treatment of obstructive lung disease, all of them target limiting exacerbations, and it is important to note that none are curative. Bronchodilators — both long- and short-acting — include beta-2-selective adrenergic agonists and anticholinergics that target airflow obstruction. The adrenergic agonists, such as albuterol and the longer-acting salmeterol, are generally inhaled and then bind and activate beta-2 adrenergic receptors of airway smooth muscle cells, leading to bronchodilation in as little as a few minutes, though the longer-acting agents have a slower time to onset [6]. Anticholinergics, such as ipratropium and tiotropium, bind muscarinic receptors of smooth muscle cells in the airways, blocking smooth muscle contraction and leading to bronchodilation. Though they bind all five subtypes of muscarinic receptors, it seems that blocking of the M3 receptor, in particular, is responsible for this bronchodilatory response [7–15].

Anti-inflammatory medications, such as corticosteroids and leukotrienes, play an important role in COPD and asthma therapy [5, 16]. The primary role of glucocorticoids is to down-regulate multiple proinflammatory pathways, resulting in an overall decrease in airway inflammation. Though they can be extremely effective in treating exacerbations through their anti-inflammatory properties, glucocorticoids are very blunt tools and come with a long list of adverse effects if administered systemically. Administration via inhalation does minimize these effects. However, while they do significantly suppress the disease associated inflammation and symptoms, they do not address the underlying process causing inflammation and, therefore, are not curative. Leukotriene inhibitors, such as montelukast, are another class of anti-inflammatory drugs that are useful in patients suffering from obstructive lung diseases, specifically asthma. Leukotrienes are a product of arachadonic acid produced by the enzyme 5-lipoxygenase that lead to slow stimulation of smooth muscle cells in the airway, causing bronchoconstriction. Leukotriene inhibitors act as antagonists at leukotriene receptors, resulting in bronchodilation and treatment of mild asthma [17, 18].

Finally, antimicrobial therapy is a mainstay of managing COPD flares. Though the exact mechanism of this therapy is not fully understood, it presumably targets local infection due to inadequate clearance contributing to the propagation of inflammation. Investigation of other drug targets, such as immunoglobulin E (IgE), has been plentiful, but relatively unsuccessful; these drugs are used only in select populations [19–26].

Each of these therapies only treats the exacerbation or manages the disease state; none is curative. Patients suffering from asthma and COPD need further investigation into potential solutions to their disease (Table 8.1).

Table 8.1 Various treatments for asthma and COPD and their action mechanisms

Disease	Treatment	Mechanism of action
COPD/ Asthma	B2-selective adrenergic agonists	Bind B-2 adrenergic receptors of airway smooth muscle cells — bronchodilation Onset of action based on specific agent
COPD	Anticholinergics	Bind muscarinic receptors of smooth muscle cells in the airways — block smooth muscle contraction M3 receptor primarily responsible
COPD/ Asthma	Corticosteroids	Down-regulate multiple proinflammatory pathways
Asthma	Leukotriene inhibitors	Inhibit products of arachadonic acid that stimulate and constrict smooth muscle cells
COPD	Antimicrobial therapy	Presumably targets the local infection due to inadequate clearance contributing to inflammation
Asthma	Cromolyn sodium	Inhibits specific chloride channels in mast cells and sensory neurons to reduce airway reactivity after exposure to inhaled irritants
Asthma	Omalizumab	Recombinant DNA-derived monoclonal antibody that selectively binds to human immunoglobulin IgE, preventing its binding to the receptor located on basophils and mast cells

8.2 Obstructive Lung Diseases

Advances in nanoparticle (NP) medicine seem particularly promising for these most common lung diseases, COPD and asthma. Both diseases' underlying pathologies involve malfunctioning of pathways in the airway immune system and in the epithelial barrier. In order to understand the ways in which COPD and asthma compromise these systems, and how NP medicine might help to treat them, it is important to review how these systems work in healthy individuals.

The airway epithelium, the lining of a pathway for gas exchange, is a critical part of the immune system, the place where the body first encounters and responds to airborne microbes and particles. The innate immune system of the airway epithelium comprises an extensive arsenal of defenses against microbial invasion. In order to decipher the pathophysiology of the obstructive lung diseases, it is essential to first understand these defenses. When a bacteria or virus enters the airway and interacts with the epithelium, the innate immune system recognizes the bacterium or virus using several different pattern recognition receptors (PRRs) that must distinguish between self and nonself.

One group of increasingly well-understood and important PRRs is the Toll-like receptors (TLRs). These receptors are expressed by the airway epithelium at cell surfaces and endosomes and serve to recognize a particular unique molecular pattern found commonly in the microbial world — from bacterial lipopolysaccharide [27] to single-stranded viral RNA to unmethylated CpG DNA [8]. TLRs serve to identify these molecular patterns and then translate that recognition into an innate immune system response. Researchers have implicated TLRs in events ranging from the recruitment of phagocytes to the site of infection to the activation of epidermal growth factor receptor (EGFR), the latter implying a role for TLRs in airway remodeling in asthma.

TLR4 is a particularly well-studied member of the TLR group of receptors. This receptor is expressed at airway epithelial cell surfaces and recognizes and binds LPS. When TLR4 binds lipopolysaccharide (LPS), it initiates the innate immune response, leading to the expression and release of cytokines such as IL-8 and IL-6. These cytokines then activate neutrophils and recruit them to the site of infection. Though this mechanism is essential to the airway's innate immune response, the activity of these phagocytes can damage the epithelial lining itself. And so the presence of a regulatory component, to ensure that the TLR4 pathway is not constitutively active, is essential. This regulation is provided by the costimulatory molecule MD2, acting together with accessory proteins CD14 and LPS-binding protein, and their intervention is essential for the successful signaling of the TLR4 pathway [28].

Both of the major obstructive lung pathologies involve these regulatory mechanisms. In asthma, which seems to have both environmental and genetic components, the house mite allergen

Der p 2 — which has been shown to trigger asthma exacerbations — has structural characteristics very similar to that of MD2. This molecular mimicry may then lead to the pathologic activation of the innate immune response via the TLR4 pathway and ultimately to airway inflammation and damage [17, 29]. TLR4 is also involved in COPD. Studies have found significantly reduced TLR4 levels in the airway epithelia of smokers with severe COPD compared to the airways of smokers with mild COPD and to those of nonsmokers. This suggests that chronic infection plays a role in COPD and COPD exacerbations [30–32].

Another important element in the epithelial immune system is secretory proteins, which are produced and released by cells of the epithelium in order to protect the airway. These proteins, which include mucins and defensins, act as the antimicrobials of the airway and also protect the epithelial barrier from desiccation.

Production of mucins can be stimulated by the presence of various types of bacteria, such as *S. aureus*, *P. aeruginosa*, and *S. pneumoniae*, or by the presence of certain viruses, such as rhinovirus and respiratory syncytial virus (RSV). Rhinovirus triggers gene expression of MUC5AC and MUC5B, leading to increased mucin production by specialized epithelial cells. This increase in mucin, via rhinovirus, has also recently been shown to trigger asthma exacerbations via increased MUC5AC production [28].

Defensins, and specifically beta-defensins (hBDs), are also produced by airway epithelial cells as part of the innate immune system's normal response. hBD2 expression, for instance, is induced by the presence of bacterial LPS and involves CD14, TLR4, and NF-KB. RSV can also stimulate hBD2 expression via a different but related pathway. The epithelial cells, when stimulated, release hBD2, and these defensins then act as antimicrobials on pathogens in the airway [33].

8.2.1 Airway Inflammation in COPD

The underlying pathophysiology of airway inflammation in COPD invokes the pathways described above. Most of the burden of COPD (~90%) can be attributed to cigarette smoking, though inhalation of small particles associated with work conditions or cooking can also contribute, especially in developing countries. Though it is unknown how smoking initially activates the innate immune system,

Matzinger's "danger hypothesis" is a likely explanation. This theory suggests that the tissue damage resulting from microbes (or in this case, xenobiotic particles and free radicals found in cigarette smoke) is the signal that provokes the innate immune system to respond (Fig. 8.1).

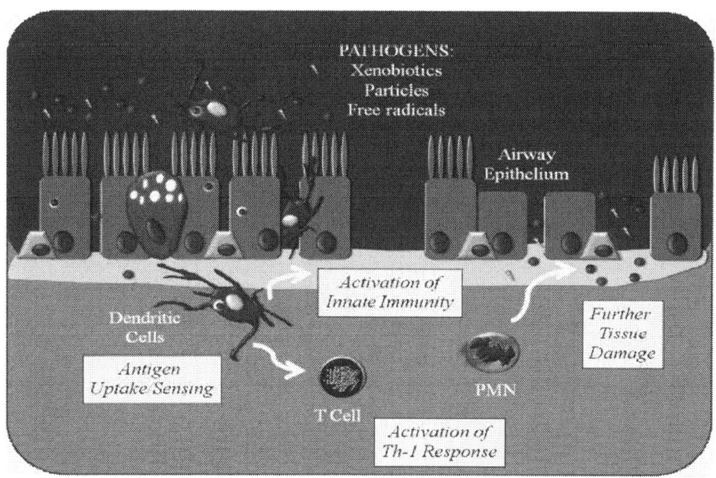

Figure 8.1 Effects of inhaled pathogens and particles on the airway epithelium and the resultant activation of the innate and adaptive immune systems. See also Color Insert.

According to this hypothesis, the immune response to this tissue damage might result in the release of autoantigens and formerly sequestered proteins that the immune system then "sees" as foreign, developing an immune response that results in further tissue damage. At this point in the development of COPD, the patient is most likely a smoker with normal lung function [34].

The next step in the evolution of COPD implicates the TLRs. Immature dendritic cells mature when their TLRs bind recognizable ligands. This initial change results in a downstream signal cascade in which affected T-cells differentiate into Type 1 helper T cells (Th1 cells), which ultimately produce interferon-gamma. Indeed, in the airways of smokers with COPD, there are increased numbers of mature dendritic cells. And in lungs of people with COPD, there are increased levels of interferon-gamma as well as Th1 cells. Increased levels of interferon-gamma correlate with worsening airflow in patients with COPD. Patients with high levels of dendritic cell stimulation of T cells correlate clinically to GOLD stage 3 or 4. Patients with increased immunoregulation of these pathways have GOLD stage 1 or 2 [34–36].

Finally, CD8+ T cells are the predominant immune cells in the airways of patients with COPD (differing from patients with asthma, who have a CD4+ T cell predominant disease). These cells can attack any cell displaying the appropriate major histocompatibility complex (MHC) class I molecule, releasing numerous proteolytic enzymes and resulting in death by apoptosis or necrosis of the target cell. These proteolytic enzymes can also contribute to the destruction of lung parenchyma seen in COPD [34–36].

8.2.2 Airway Inflammation in Asthma

Asthma has an overlapping but different underlying pathophysiology. The initial insult is thought to usually come from either a virus or an environmental allergen — or both — which triggers a pathologic immune response. The "hygiene hypothesis" suggests that early exposure to microbiologic agents triggers an initial predominant Th1 type response instead of the asthmatic-phenotype Th2-predominant response — though as researchers understand more about the pathogenesis of asthma, they suspect that this may overly simplify the process. Even so, it is clear that asthma represents a distortion of the normal airway immune response.

One of the pathways implicated in airway inflammation involves P-selectin, an epithelial adhesion molecule. Early in the event of the immune response sequence, P-selectins recruit leukocytes to the area of infection, leading to further inflammation [37]. P-selectin has been implicated in asthma and COPD and has been identified as a potential therapeutic target.

8.2.3 Role of the Airway Epithelial Barrier

Both COPD and asthma involve imbalances in not only the epithelial innate immune system but also the epithelial barrier. The airway epithelium serves as the body's interface with the outside world, and so one of its primary roles is to gate the interaction between the lumen of the airway and subepithelial tissues. Increases in airway epithelial paracellular permeability occur in a variety of circumstances, including airway infection or inflammation and asthma, and that disruption is believed to contribute to disease expression. The regulation of this epithelial barrier depends on a complex of proteins that compose distinct intercellular junctions — tight junctions, adherens junctions, and desmosomes — each of which exhibits cell type–specific regulation by various growth factors, agonists, and second messengers [38–42].

The airway epithelium serves two distinct barrier roles, (1) as a physical barrier that regulates the access of luminal contents to regions located beneath the apical membrane of the epithelial cells and (2) as an immunologic barrier, in which the epithelial cells provide the first defense against the infectious and noninfectious respirable particles that are in the airstream. Some research suggests that these two roles are related and that by regulating the physical barrier, the epithelial cells can modify the innate immunologic response. For example, histamine increases epithelial permeability and this effect is believed to be linked to the increased airway responsiveness and the increased inflammation associated with histamine release, though the mechanisms mediating this pathway are not known [43].

Using NP medicine, we may be able to target the airway epithelium to alter the barrier at the first sight of injury and provide additional therapies for obstructive lung diseases.

8.2.3.1 Epithelial barrier in asthma

Researchers have known for some time that the airway epithelial barrier function is altered in asthmatics [15–18], though the mechanisms mediating this alteration are not known. Data obtained from biopsies from asthmatic children suggests that the airway epithelium has increased paracellular permeability and decreased membrane expression of those proteins that mediate cell-cell contacts, such as ZO-1. Investigators do not know whether this is a marker of diseased epithelium or whether it is integral to the pathogenesis. Research has shown that altered barrier function is found early in disease pathogenesis and is not merely a consequence of chronic inflammation [44–49]. In addition, some research has suggested that these changes persist *in vitro* when cells from asthmatic airways are cultured. These innate changes in the epithelium are likely fundamental to understanding the body's response to allergens.

None of the therapies currently used in the treatment of obstructive lung disease target the epithelial barrier. And while corticosteroids effectively control many of the inflammatory parameters of asthma, they have no known effect on the epithelial barrier function [50–55]. Targeting the epithelium is a potential new approach to asthma and COPD therapy and may provide options for curative treatments that currently do not exist [56]. Such therapies might also offer insight into diseases involving epithelial barriers elsewhere in the body — for instance, in the gut.

8.2.3.2 Epithelial barrier in COPD

As in asthma, the epithelial barrier in COPD is altered so as to increase its permeability [57]. Cigarette smoke, the most common cause of COPD, directly damages the epithelium through the cytotoxic effects of certain components of the smoke. Cigarette smoke has also been shown to have a direct effect on gene expression of the apical junction complex, again leading to increased permeability of the epithelial barrier [57].

8.3 NP Delivery of Airway Diseases

8.3.1 Local Delivery

NP delivery of drugs for chronic obstructive lung disease is an attractive option given the ease of local inhalational delivery. This targeted delivery can potentially result in reducing the overall dose and the amount of side effects that result from high levels of systemic drug exposure. In a mouse model of pulmonary aspergillosis, investigators compared the efficacy of inhalational nanostructured itroconazole delivered by nebulizers to that of the oral delivery of liquid itroconazole and found that in the nanodelivery system, there was improved survival and limited invasive disease of small airways [58]. This suggests that exploring inhalational and nanodelivery options has merit. The pulmonary delivery of therapeutic NPs has been explored for a variety of molecules with a range of sizes from 150 nm to 500 nm and different delivery mechanisms. The main attraction of using submicron-sized particles in lung delivery, particularly those under 260 nm in size, is the observation that such small particles tend to escape the detection systems of alveolar macrophages and remain in the lung long enough to release their contents in a controlled manner [59].

However, along with the notable benefits of the local NP delivery of drugs come some limitations. NPs that are too small in size are cleared with exhalation without sufficient drug delivery, though determination of what this size is depends on the mode of administration — instillation versus inhalation [60] — as well as the charge of the particle [61]. For example, some studies using instillation as a techniques show that there is a high concentration of particle traversing into the bloodstream when they are <6 nm and zwitterionic [61] while others suggest that there is no deposition of particles that small [62, 63]. Studies have suggested that there is increased laryngeal deposition of micron-sized particles (1–10 μm), bronchial deposition of particles 10–50 nm in size, and alveolar deposition of particles 50–100 nm in size (Fig. 8.2), but as mentioned above, that is largely influenced by the charge and mode of delivery [63].

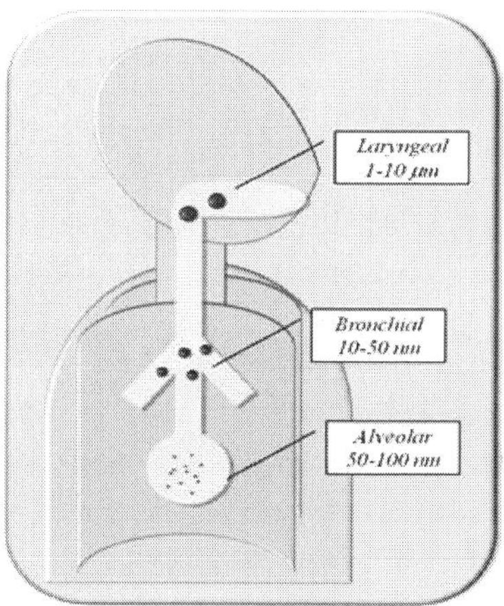

Figure 8.2 The distribution of the deposition of nanoparticles based on size, assuming neutral charge. See also Color Insert.

On the other end of the spectrum, with particles that do get deposited in the airways, there is concern about particle fate after delivery of the drug. It has been noted that some particles remain in the body either locally in the lung or with absorption into system tissues such as liver and brain for long after the drug has been delivered. Choi *et al.* studied the fate of the particle once it came into contact with the lung to assess some of these issues. Their studies suggested that a size threshold of ~34 nm determines whether there is rapid transepithelial translocation of NPs from the alveolar luminal surface into the septal interstitium. Below this size, surface charge is a major factor that influences translocation, with zwitterionic, anionic, and polar surfaces being permissive and cationic surfaces being restrictive to translocation into and past the lungs. Finally, they found that zwitterionic particles <6 nm can enter the bloodstream quickly from the alveolar airspaces and can ultimately be cleared by renal filtration [61].

There is potential for significant local inflammatory effects of the lingering NP carrier. These concerns have led to the development of biodegradeable NPs comprising many materials, with the hopes that these particles will be degraded by the cells before inducing inflammatory effects.

8.3.2 Existing Studies for the Treatment of Chronic Airway Diseases

Because the major delivery system for obstructive lung diseases, both asthma and COPD, is aerosolic, NPs seem particularly well suited for treatment of these diseases. Drugs or biologically active molecules can be delivered by NPs through a host of mechanisms. They can be captured within the particle core, either by being dissolved or encapsulated in NPs. Or they can be attached to the NP by being adsorbed or chemically attached to the surface [64, 65]. The decision of how best to couple the two is based on the physicochemical properties of the carrier and the drug or bioactive agent. In terms of the particle itself, there are two types of possible inner structures. One is a matrix type, with oligomer or polymer units, and is termed a nanosphere. The other consists of a hydrophobic core surrounded by a capsule, termed a nanocapsule. Finally, lipids can also be used to form NP carriers, and these are called micelles or liposomes [64, 65]. Using nanotechnology, a P-selectin antagonist in a murine model of allergic asthma has been studied for its anti-inflammatory therapeutic potential. P-selectin is known to play an important role in peribronchial inflammation in asthma, and it was thought that a P-selectin inhibitor would have strong therapeutic potential. Using a polyvalent polymer NP that displayed multiple low-molecular-weight ligands, each with only low affinity for P-selectin, John *et al.* developed a construct with very high affinity for P-selectin. They then demonstrated that these synthetic NPs displaying P-selectin blocking arrays reduced peribronchial airway inflammation *in vivo* in a murine model of allergic asthma [13].

Another study examined chitosan/interferon (IFN-γ) pDNA NPs in a murine model of allergic asthma. Chitosan, or depolymerized chitosan oligomers, can be coupled with a variety of plasmids to form nanospheres <200 nm [64, 66, 67]. IFN-γ has been examined as a potential therapy for asthma because of its role in promoting T-helper type 1 responses that decrease Th2-type responses, which are primarily responsible for allergic reactions such as those seen in asthma. However, because of its very short half-life, IFN-γ alone is not a feasible therapeutic. Adenovirus-delivered gene therapy of IFN-γ is also untenable because of the immune response generated by the adenovirus itself. Chitosan is a compact, nonimmunogenic molecule derived from the shells of crustaceans. The authors linked chitosan and IFN-γ pDNA together in NPs and studied its potential. Female wildtype and STAT4 -/- BALB/c ovalbumin-sensitized mice were given 25 μg of intranasal Chitosan/IFN-γ NPs. The mice treated with chitosan/IFN-γ NP had significantly lower airway hyperresponsiveness to methacholine challenge when compared to placebo-treated mice. The experiment was supported by evidence on the molecular level as well: NP-treated mice had an increased level of IFN-γ and a decrease in the levels of Th2 cytokines, IL-4, IL-5, and ovalbumin-related IgE [31].

RSV is the leading cause of infant pneumonia and severe bronchiolitis and is a major risk factor for the development of asthma. However, no vaccine or other prophylaxis is currently available to prevent RSV. Kumar *et al.* investigated the use of chitosan nanospheres containing an assortment of plasmid DNA containing all RSV antigens except L in the prevention of RSV in mice. Using a similar strategy to that described above, they created chitosan nanospheres containing RSV plasmid DNA. They then intranasally administered 25 μg of the nanospheres to BALB/c mice and 16 days after the vaccination, infected the mice with RSV. The investigators found that the immunized mice did express the RSV antigens. Mice treated with nanospheres had a 100-fold reduction in RSV titers compared to control mice and also had a significant reduction in RSV-induced pulmonary inflammation. In order to assess whether the treated mice developed specific anti-RSV immune responses after inoculation with nanospheres, the investigators measured RSV-specific mucosal IgA as well as serum IgG. They found that treated mice had developed both a systemic and mucosal response. Finally, the authors found that treated mice developed specific cytotoxic T-lymphocytic responses to RSV and also had high levels of IFN-γ, the major antiviral cytokine [32].

8.4 Toxicity of NPs

Though the therapeutic and imaging value of NPs are extremely promising, they also present a paradox, in that NPs, especially those produced by combustion, are thought to be responsible for most of the human diseases attributed to air pollution. Air pollution is known to be related to lung diseases — notably, asthma, and COPD — as well as cardiovascular diseases and lung cancer. If NPs are to be used increasingly in modern medicine, we will need to understand their toxicity and how to minimize it while maximizing the beneficial properties NPs can offer.

Combustion-derived nanoparticles (CDNPs) are inorganic, carbon-centered particles with surface-associated metals, inorganics, and organics that are the products of industrial processes but mostly of automotive combustion. In urban areas, especially, they are inhaled in large quantities and deposited in the lungs, the skin, and the gut where they accumulate and cause disease. CDNPs have been studied for years by particle toxicologists aiming to understand the mechanisms of pollution-related diseases. Much of the interest in studying CDNPs and their properties as they relate to immunogenicity and disease stems from the PM10 hypothesis, which suggests that the adverse effects of PM10 are related mostly to the toxicity of NPs. (PM10 is the globally accepted convention that describes particulate matter [PM] less than 10 micrometers in diameter, particles that are likely responsible for the burden of disease — pulmonary and cardiovascular — associated with air pollution.) PM10 contains a range of particle sizes, both particulate matter larger than NPs as well as NPs themselves.

A study of six US cities showed that an increased concentration of fine particulate matter (PM2.5) correlated best with increased mortality. And a German study examining asthmatic exacerbations in response to a high pollution load found that increases in exacerbations were best associated with higher concentrations of ultrafine particles (or NPs), and not with coarser materials.

All NPs, though, not just CDNPs, induce inflammation. Though the materials making up NPs differ (they share only size), a common mechanism of oxidative stress leading eventually to cell injury likely exists. Compounds that are nontoxic (or less toxic) in fine particle size may be toxic in NP size. For instance, fine carbon black (or carbon black with a larger particle size) produces very little inflammation when dosed in mice. However, ultrafine carbon black (<100 nm) produces significant inflammation in the same setting [68–70]. In fact, in experiments done by Donaldson *et al.*, researchers found that ultrafine carbon black, a substance known to be benign in larger particle size, produced significantly more inflammation, as measured by neutrophils influx into mouse lungs, than an equivalent dose of PM10 [71]. The same group found a linear relationship between the surface area of instilled particles and the amount of inflammation caused (i.e., smaller particles produced progressively more inflammation) [68].

Interestingly, titanium dioxide (TiO_2), a relatively inert substance in bulk used as a control dust in air pollution studies, can also exert harmful immunogenic effects when administered in NP size. Ferin *et al.* demonstrated this when they administered equal masses of TiO_2 in fine (PM10) and ultrafine (nanoscale) particles, or UFPs, to rats and saw that the UFPs induced a much greater level of lung inflammation [72].

A large surface area is likely a primary characteristic driving the immunogenic effects of NPs — NPs have a high surface-to-mass ratio compared to larger particles of the same materials. For instance, Donaldson *et al.* describe that carbon black has a surface area of 7.9 m^2/g while ultrafine carbon black has a surface area of 253.9 m^2/g, or carbon black has 32.1 times more surface area per unit mass than ultrafine carbon black. This large surface area — and therefore large potential for interaction with tissues and cells compared to their non-NP counterparts — makes them uniquely useful in medicine as well as potentially dangerous [68]. Additional factors involved in the potential for NPs to cause oxidative stress are related to their shape and surface charge [68].

An important difference in the CDNPs that lead to pollution -related disease and the medical NPs is immunogenicity of the material, even at NP size [73–75]. CDNPs are inorganic, carbon-centered particles recognized by the immune system as foreign. Medical NPs are usually made from biocompatible materials that do not generate an immune response, such as chitosan and polyethylene glycol (PEG). These do seem to lack the strong immunogenicity of their CDNP counterparts, but further study is needed to examine whether these materials are indeed safe for medical use.

Clearly, as we move forward in understanding the potential benefits of NPs in therapeutics and imaging, we must further our understanding of the potential toxicity of these molecules. Currently, there is no standard approach to the evaluation of toxicity of a newly manufactured NP. As NP use increases, a more systematic approach to the study of their toxicity will be important.

8.5 Conclusions and Future Directions

COPD and asthma together cause a significant portion of global morbidity and mortality, leading to more than 3 million deaths worldwide each year and affecting more than 510 million people [1, 2]. By 2020, they are predicted to become the third-leading cause of death in the United States. However, treatments for obstructive lung diseases are limited to preventing and controlling exacerbations. Nothing available currently is curative of either COPD or asthma, and no currently identified therapy stops the progression of COPD.

NP therapy — delivered via aerosol directly to the airway epithelium — seems particularly well suited to treat these diseases as it allows for local, targeted therapy. In addition, it is a wide-open, new field of opportunity for therapy development. Currently, researchers are approaching NP therapies in obstructive lung diseases from a number of angles. Studies have targeted P-selectin via NP-coupled-P-selectin inhibitors [37]. Chitosan-linked therapies have also been of particular interest. In addition, linking immunomodulatory therapy such as IFN-γ to a nonimmunogenic NP, chitosan has provided considerable interest in the field of asthma and COPD therapy [64, 74].

However, as we have also discussed, NP medicine is not without risk. Research investigated NPs for their potential dangers, specifically in air pollution, long before their possible therapeutic benefits were known, and further studies are needed to determine how best to minimize that risk, be it size/shape or charge manipulation or the specifics of targeting subsequent clearance of the NPs.

The use of NP medicine in the treatment of asthma and COPD provides an exciting opportunity to address this important group of diseases in previously impossible ways. Currently, however, research into potential NP therapies for COPD is scant. A PubMed search for "COPD and nanoparticle" does not retrieve a single original research paper in this area. COPD-directed NP therapy could potentially have the ability to address complex issues such as mucus hypersecretion by molecular-manipulating mucin gene expression.

Targeting mucin production — via NP-coupled inhibitors of protein expression, whether by molecularly targeting transcription and translation or accelerating protein degradation — would be a way to address hypersecretion in airways of both COPD and asthma patients. The resulting decrease in airway mucus production could lead to a reduction in exacerbations and possibly prevention of further airway remodeling.

In addition to targeting mucins, other molecules involved in this pathway, such as IL-1 beta, TNF-alpha, and cyclooxygenase-2, could be addressed. A more downstream, proinflammatory target might be EGFR, an activator of airway inflammation in asthma and COPD. Finally, elastases or other tissue-destructive mechanisms that lead to the development of emphysema could be a target of NP-coupled therapy. By blocking elastases, an NP-based therapy might prevent these enzymes' further destruction of lung tissue in COPD. To date, there have been no therapies addressing these targets.

Besides providing a vehicle for therapeutic delivery, the use of NPs in research will provide us with a unique opportunity to study the effects of specific proteins known to contribute to disease manifestation on the disease process, as at this point we do not completely understand all the roles these proteins have in normal and diseased airways. However, to allow us to use this as an appropriate research and therapeutic tool, we must first understand how best to minimize toxicity. As NP medicine moves forward, it will be essential to address the potential for NP-specific toxicity similar to that seen in air pollution. And, therefore, a more robust system for the assessment of safety is necessary to pursue the research and therapeutic use of nanomedicine.

References

1. WHO, *Asthma* [Internet]. 2010.
2. WHO, *Chronic obstructive pulmonary disease (COPD)* [Internet]. 2010.
3. CDC, *FASTSTATS — Leading Causes of Death* [Internet]. 2010.
4. AAAAI, *Statistics* [Internet]. 2010.
5. Pauwels, R., Global initiative for chronic obstructive lung diseases (GOLD): Time to act. *Eur Respir J*, 2001. **18**(6): 901–902.
6. Westfall, T.C., and D.P. Westfall (2006) Adrenergic agonists and antagonists, in *Goodman & Gilman's the Pharmacological Basis of Therapeutics* (ed. Lazo, J.S., Brunton, L.L., and Parker, K.L.), McGraw-Hill, New York, pp. 237–295.
7. Campbell, S.C., Clinical aspects of inhaled anticholinergic therapy. *Respir Care*, 2000. **45**(7): 864–867.
8. Decramer, M. *et al.*, Clinical trial design considerations in assessing long-term functional impacts of tiotropium in COPD: the UPLIFT trial. *COPD*, 2004. **1**(2): 303–312.
9. Groeben, H., and R.H. Brown, Ipratropium decreases airway size in dogs by preferential M2 muscarinic receptor blockade in vivo. *Anesthesiology*, 1996. **85**(4): 867–873.
10. Hansel, T.T., and P.J. Barnes, Tiotropium bromide: a novel once-daily anticholinergic bronchodilator for the treatment of COPD. *Drugs Today (Barc)*, 2002. **38**(9): 585–600.
11. Howell, R.E. *et al.*, Pulmonary pharmacology of a novel, smooth muscle-selective muscarinic antagonist in vivo. *J Pharmacol Exp Ther*, 1994. **270**(2): 546–553.
12. Mitsuya, M. *et al.*, J-104129, a novel muscarinic M3 receptor antagonist with high selectivity for M3 over M2 receptors. *Bioorg Med Chem*, 1999. **7**(11): 2555–2567.
13. Proskocil, B.J., and A.D. Fryer, Beta2-agonist and anticholinergic drugs in the treatment of lung disease. *Proc Am Thorac Soc*, 2005. **2**(4): 305–310; discussion 311–312.
14. Roux, E. *et al.*, Muscarinic stimulation of airway smooth muscle cells. *Gen Pharmacol*, 1998. **31**(3): 349–356.
15. Shioya, T. *et al.*, Antimuscarinic effect of tiquizium bromide in vitro and in vivo. *Eur J Clin Pharmacol*, 1996. **50**(5): 375–380.
16. Kips, J.C., and R.A. Pauwels, Low dose inhaled corticosteroids and the prevention of death from asthma. *Thorax*, 2001. **56 Suppl 2**: ii74–ii78.

17. O'Byrne, P.M., G.M. Gauvreau, and D.M. Murphy, Efficacy of leukotriene receptor antagonists and synthesis inhibitors in asthma. *J Allergy Clin Immunol*, 2009. **124**(3): 397–403.

18. Larsen, J.S., and E.P. Acosta, Leukotriene-receptor antagonists and 5-lipoxygenase inhibitors in asthma. *Ann Pharmacother*, 1993. **27**(7–8): 898–903.

19. D'Amato, G. *et al.*, Treating moderate-to-severe allergic asthma with anti-IgE monoclonal antibody (omalizumab). An update. *Eur Ann Allergy Clin Immunol*, **42**(4): 135–140.

20. Fried, A.J., and H.C. Oettgen, Anti-IgE in the treatment of allergic disorders in pediatrics. *Curr Opin Pediatr*, 2010. **22**(6): 758–764.

21. Rodrigo, G.J., H. Neffen, and J.A. Castro-Rodriguez, Efficacy and safety of subcutaneous omalizumab versus placebo as add on therapy to corticosteroids for children and adults with asthma: a systematic review. *Chest*, 2011. **139**(1): 28–35.

22. Levine, S.J., and S.E. Wenzel, Narrative review: the role of Th2 immune pathway modulation in the treatment of severe asthma and its phenotypes. *Ann Intern Med*, **152**(4): 232–237.

23. Lanier, B. *et al.*, Omalizumab for the treatment of exacerbations in children with inadequately controlled allergic (IgE-mediated) asthma. *J Allergy Clin Immunol*, 2009. **124**(6): 1210–1216.

24. Corren, J. *et al.*, Safety and tolerability of omalizumab. *Clin Exp Allergy*, 2009. **39**(6): 788–797.

25. Peng, Z., Vaccines targeting IgE in the treatment of asthma and allergy. *Hum Vaccin*, 2009. **5**(5): 302–309.

26. Pelaia, G. *et al.*, Omalizumab in the treatment of severe asthma: efficacy and current problems. *Ther Adv Respir Dis*, 2008. **2**(6): 409–421.

27. Jia, Y. *et al.*, Functional TRPV4 channels are expressed in human airway smooth muscle cells. *Am J Physiol Lung Cell Mol Physiol*, 2004. **287**(2): L272– L278.

28. Ryu, J.H., C.H. Kim, and J.H. Yoon, Innate immune responses of the airway epithelium. *Mol Cells*, **30**(3): 173–183.

29. Murphy, D.M., and P.M. O'Byrne, Recent advances in the pathophysiology of asthma. *Chest*, **137**(6): 1417–1426.

30. Kelly, E. *et al.*, Community-acquired pneumonia in older patients: does age influence systemic cytokine levels in community-acquired pneumonia? *Respirology*, 2009. **14**(2): 210–216.

31. MacRedmond, R. *et al.*, Respiratory epithelial cells require Toll-like receptor 4 for induction of human beta-defensin 2 by lipopolysaccharide. *Respir Res*, 2005. **6**: 116.

32. MacRedmond, R.E. *et al.*, Epithelial expression of TLR4 is modulated in COPD and by steroids, salmeterol and cigarette smoke. *Respir Res*, 2007. **8**: 84.
33. Tecle, T., S. Tripathi, and K.L. Hartshorn, Review: defensins and cathelicidins in lung immunity. *Innate Immun*, **16**(3): 151–159.
34. Cosio, M.G., M. Saetta, and A. Agusti, Immunologic aspects of chronic obstructive pulmonary disease. *N Engl J Med*, 2009. **360**(23): 2445–2454.
35. Baraldo, S., M. Saetta, and M.G. Cosio, Pathophysiology of the small airways. *Semin Respir Crit Care Med*, 2003. **24**(5): 465–472.
36. Saetta, M., R. Finkelstein, and M.G. Cosio, Morphological and cellular basis for airflow limitation in smokers. *Eur Respir J*, 1994. **7**(8): 1505–1515.
37. John, A.E. *et al.*, Discovery of a potent nanoparticle P-selectin antagonist with anti-inflammatory effects in allergic airway disease. *Faseb J*, 2003. **17**(15): 2296–2298.
38. Anderson, J.M., and C.M. Van Itallie, Tight junctions and the molecular basis for regulation of paracellular permeability. *Am J Physiol*, 1995. **269**(4 Pt 1): G467–G475.
39. Anderson, J.M., and C.M. Van Itallie, Tight junctions: closing in on the seal. *Curr Biol*, 1999. **9**(24): R922– R924.
40. Anderson, J.M., and C.M. Van Itallie, Tight junctions. *Curr Biol*, 2008. **18**(20): R941– R943.
41. Anderson, J.M., and C.M. Van Itallie, Physiology and function of the tight junction. *Cold Spring Harbor Perspect Biol*, 2009. **1**(2): a002584.
42. Anderson, J.M., C.M. Van Itallie, and A.S. Fanning, Setting up a selective barrier at the apical junction complex. *Curr Opin Cell Biol*, 2004. **16**(2): 140–145.
43. Zabner, J. *et al.*, Histamine alters E-cadherin cell adhesion to increase human airway epithelial permeability. *J Appl Physiol*, 2003. **95**(1): 394–401.
44. Fedorov, I.A. *et al.*, Epithelial stress and structural remodelling in childhood asthma. *Thorax*, 2005. **60**(5): 389–394.
45. Hamilton, L.M. *et al.*, The bronchial epithelium in asthma--much more than a passive barrier. *Monaldi Arch Chest Dis*, 2001. **56**(1): 48–54.
46. Hamilton, L.M. *et al.*, Altered protein tyrosine phosphorylation in asthmatic bronchial epithelium. *Eur Respir J*, 2005. **25**(6): 978–985.

47. Hamilton, L.M. *et al.*, The role of the epidermal growth factor receptor in sustaining neutrophil inflammation in severe asthma. *Clin Exp Allergy*, 2003. **33**(2): 233–240.

48. Holgate, S.T. *et al.*, Epithelial-mesenchymal interactions in the pathogenesis of asthma. *J Allergy Clin Immunol*, 2000. **105**(2 Pt 1): 193–204.

49. Holgate, S.T. *et al.*, Epithelial-mesenchymal communication in the pathogenesis of chronic asthma. *Proc Am Thorac Soc*, 2004. **1**(2): 93–98.

50. Djukanovic, R. *et al.*, Bronchial mucosal manifestations of atopy: a comparison of markers of inflammation between atopic asthmatics, atopic nonasthmatics and healthy controls. *Eur Respir J*, 1992. **5**(5): 538–544.

51. Djukanovic, R. *et al.*, Mucosal inflammation in asthma. *Am Rev Respir Dis*, 1990. **142**(2): 434–457.

52. Djukanovic, R. *et al.*, Effect of an inhaled corticosteroid on airway inflammation and symptoms in asthma. *Am Rev Respir Dis*, 1992. **145**(3): 669–674.

53. Holgate, S.T. *et al.*, Allergic inflammation and its pharmacological modulation in asthma. *Int Arch Allergy Appl Immunol*, 1991. **94**(1–4): 210–217.

54. Holgate, S.T. *et al.*, Inflammatory processes and bronchial hyper-responsiveness. *Clin Exp Allergy*, 1991. **21 Suppl 1**: 30–36.

55. Holgate, S.T. *et al.*, The need for a pathological classification of asthma. *Eur Respir J Suppl*, 1991. **13**: 113s–122s.

56. Boucher, R.C. *et al.*, The effect of cigarette smoke on the permeability of guinea pig airways. *Lab Invest*, 1980. **43**(1): 94–100.

57. Shaykhiev, R. *et al.*, Cigarette smoking reprograms apical junctional complex molecular architecture in the human airway epithelium in vivo. *Cell Mol Life Sci*, 2011. **68**(5): 877–892.

58. Alvarez, C.A. *et al.*, Aerosolized nanostructured itraconazole as prophylaxis against invasive pulmonary aspergillosis. *J Infect*, 2007. **55**(1): 68–74.

59. Niven, R.W., Delivery of biotherapeutics by inhalation aerosol. *Crit Rev Ther Drug Carrier Syst*, 1995. **12**(2–3): 151–231.

60. Mossman, B.T. *et al.*, Mechanisms of action of inhaled fibers, particles and nanoparticles in lung and cardiovascular diseases. *Part Fibre Toxicol*, 2007. **4**: 4.

61. Choi, H.S. *et al.*, Rapid translocation of nanoparticles from the lung airspaces to the body. *Nat Biotechnol*, **28**(12): 1300–1303.

62. Rogueda, P.G., and D. Traini, The nanoscale in pulmonary delivery. Part 2: formulation platforms. *Expert Opin Drug Deliv*, 2007. **4**(6): 607–620.

63. Rogueda, P.G., and D. Traini, The nanoscale in pulmonary delivery. Part 1: deposition, fate, toxicology and effects. *Expert Opin Drug Deliv*, 2007. **4**(6): 595–606.

64. Ravi Kumar, M. *et al.*, Nanoparticle-mediated gene delivery: state of the art. *Expert Opin Biol Ther*, 2004. **4**(8): 1213–1224.

65. Ravi Kumar, M.N. *et al.*, Cationic silica nanoparticles as gene carriers: synthesis, characterization and transfection efficiency in vitro and in vivo. *J Nanosci Nanotechnol*, 2004. **4**(7): 876–881.

66. Pulliam, B., J.C. Sung, and D.A. Edwards, Design of nanoparticle-based dry powder pulmonary vaccines. *Expert Opin Drug Deliv*, 2007. **4**(6): 651–663.

67. Sung, J.C., B.L. Pulliam, and D.A. Edwards, Nanoparticles for drug delivery to the lungs. *Trends Biotechnol*, 2007. **25**(12): 563–570.

68. Donaldson, K. *et al.*, The pulmonary toxicology of ultrafine particles. *J Aerosol Med*, 2002. **15**(2): 213–220.

69. Li, X.Y. *et al.*, Short-term inflammatory responses following intra-tracheal instillation of fine and ultrafine carbon black in rats. *Inhal Toxicol*, 1999. **11**(8): 709–731.

70. Stone, V. *et al.*, The role of oxidative stress in the prolonged inhibitory effect of ultrafine carbon black on epithelial cell function. *Toxicol In Vitro*, 1998. **12**(6): 649–659.

71. Renwick, L.C. *et al.*, Increased inflammation and altered macrophage chemotactic responses caused by two ultrafine particle types. *Occup Environ Med*, 2004. **61**(5): 442–447.

72. Ferin, J., G. Oberdorster, and D.P. Penney, Pulmonary retention of ultrafine and fine particles in rats. *Am J Respir Cell Mol Biol*, 1992. **6**(5): 535–542.

73. Semete, B. *et al.*, In vivo evaluation of the biodistribution and safety of PLGA nanoparticles as drug delivery systems. *Nanomedicine*, **6**(5): 662–671.

74. Semete, B. *et al.*, In vivo uptake and acute immune response to orally administered chitosan and PEG coated PLGA nanoparticles. *Toxicol Appl Pharmacol.* **249**(2): 158–165.

75. Swai, H. *et al.*, Nanomedicine for respiratory diseases. *Wiley Interdiscip Rev Nanomed Nanobiotechnol*, 2009. **1**(3): 255–263.

Chapter 9

Potential Respiratory Health Risks of Engineered Carbon Nanotubes

James C. Bonner,[a,*] Jeffrey W. Card,[b] Stavros Garantziotis,[c] and Darryl C. Zeldin[d]

[a]*Department of Environmental and Molecular Toxicology, North Carolina State University, Raleigh, NC 27695, USA*
[b]*Ashuren Health Sciences, Mississauga, Ontario, Canada*
[c, d]*Laboratory of Respiratory Biology, NIH/NIEHS, Research Triangle Park, NC 27709, USA*
[*]james_bonner@ncsu.edu

9.1 Introduction

Nanotechnology offers great potential benefits for drug delivery and therapy of respiratory diseases, including asthma, chronic obstructive pulmonary disease (COPD), and pulmonary fibrosis. Nanoparticles (NPs) have been of great interest for some time as they can be designed to simultaneously carry a drug payload, specifically target features of diseased tissues, and carry an imaging molecule to track drug accumulation and clearance in tissues. Moreover, they can be engineered for more sustained drug delivery to improve pharmacokinetics. A variety of NPs have been investigated in experimental animal models as tools to improve the therapeutic

Pulmonary Nanomedicine: Diagnostics, Imaging, and Therapeutics
Edited by Neeraj Vij

ISBN 978-981-4316-48-4 (Hardback), 978-981-4364-14-0 (eBook)
www.panstanford.com

efficacy of drugs or genes delivered to the lung or other organ systems [1]. The nanotechnology platform for drug delivery contains a number of very different types of nanostructures with widely varying properties, including dendrimers, fullerenes, carbon nanotubes (CNTs), and polymeric NPs. While nano-based drug delivery systems offer the potential for improved the efficacy of treatment, there are also potential risks associated with these novel therapeutic strategies [2]. Engineered NPs are desirable because they easily enter cells and they can be designed to interact with specific cellular structures (e.g., receptors) to allow for accumulation within specific regions of the cell. NPs can also be designed as pH-labile structures so that they degrade within the more acidic microenvironment of the cell to release drug payloads. Biocompatible NPs that are used for drug delivery, such as poly-(ethylene glycol)-*block*-lactide/glycolide copolymer (PEG-PLGA) NPs, generally have low toxicity but often do not persist in tissues long enough for sustained drug or gene payload delivery. Nevertheless, PEG-PLGA NPs might be useful for delivering metabolic inhibitors of inflammation for the treatment of pulmonary hypertension [3]. CNTs are durable and persist in biological systems for weeks or longer. Moreover, their tube- or fiberlike structure allows for extensive functionalization and loading of cargo. The versatile physicochemical features of CNTs allow for covalent and noncovalent functionalization to simultaneously carry several agents: (1) a drug, (2) an imaging agent (to track the course of delivery), and (3) a specific targeting agent (e.g., the antibody selective for the diseased tissue) [4]. Despite the potential benefits of CNTs for drug delivery, some of the same unique properties that make CNTs desirable for therapeutic applications also make them potentially toxic. Thus, despite potential therapeutic value for certain diseases, the toxic effects of CNTs might ultimately outweigh any beneficial effects as drug delivery systems. While no information yet exists on the adverse effects of CNTs in humans, growing evidence from *in vivo* exposure studies with rodents and *in vitro* cell culture models demonstrates that CNTs cause numerous adverse effects, including lung fibrosis, exacerbation of preexisting lung disease, adverse immune reactions, and DNA damage. The aims of this chapter is to provide an overview of some of the evidence that shows that CNTs have potential adverse health effects and analyze why CNTs should be carefully evaluated and designed if they are to be pursued further as drug delivery platforms.

9.2 Immune Cell Interaction with CNTs

Macrophages are the first line of immune defense to inhaled particles. Some studies indicate that single-walled CNTs (SWCNTs) delivered to the lungs of mice escape immune surveillance by macrophages in that they are not readily recognized and taken up by phagocytosis unless they are treated with agents such as phosphatidylserine [5]. However, other studies have shown that SWCNTs are readily recognized and avidly engulfed by macrophages. Work by Mangum and colleagues showed the uptake of SWNCTs by rat alveolar macrophages *in vivo* and further showed the unusual formation of SWCNT bridge structures that linked two macrophages after 21 days of exposure [6]. These bridge structures were not observed in rats exposed to 8 nm diameter carbon black NPs, indicating that the carbon bridge structures between macrophages were unique to CNTs (Fig. 9.1). Furthermore, the nature of carbon bridge formation remains to be elucidated, but it is conceivable that the nanoscale

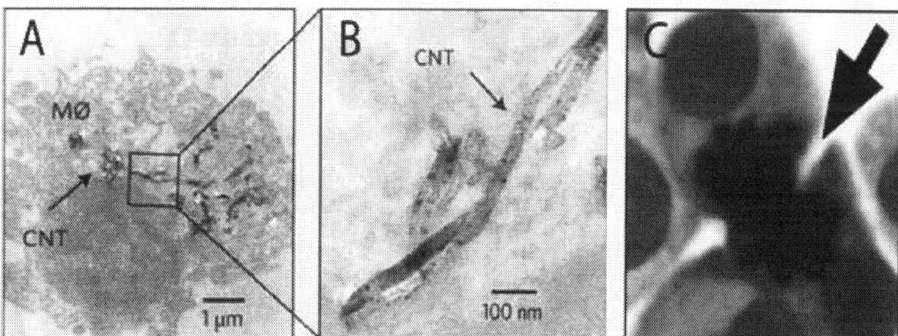

Figure 9.1 Macrophage interactions with multiwalled carbon nanotubes (MWCNTs) and single-walled carbon nanotubes (SWCNTs). Panel A shows a TEM image of MWCNT engulfed by a macrophage within the lungs of a mouse after inhalation exposure (Panel B shows detail). Panel C shows a light microscope image of two macrophages joined by a bridge structure composed of SWCNTs [6]. See also Color Insert.

width of CNTs might allow for interaction with subcellular structures such as cytoskeletal filaments. Interestingly, actin filaments that comprise the cytoskeleton are approximately the same diameter as SWCNT, and therefore carbon bridge structures might represent a molecular interaction between SWCNTs and actin. Whether or not SWCNTs are recognized by macrophages may depend on the aggregation state of the CNT preparation. It is likely that more dispersed SWCNTs would not be readily recognized whereas aggregates of SWCNT would be recognized as larger particles and taken up by macrophages. Therefore, the aggregation status is likely an important determinant of evaluating the toxicity of CNTs. CNTs that are engulfed by macrophages are removed from the lung through two principal avenues — the mucociliary escalator and the pulmonary lymphatic system. The mucociliary escalator transports macrophages and their engulfed payload up the airways of the lungs, where they are ultimately swallowed or expelled through coughing. The clearance of CNT-laden macrophages through the pulmonary lymphatic system could lead to adverse effects as many of these macrophages traffic across the pleura region, where it is thought that CNTs might have potential carcinogenicity similar to asbestos fibers [7].

9.3 Fibrogenic Reactions to CNTs

CNTs have attracted the attention of respiratory toxicologists due to their fiberlike shape and high aspect (length to width) ratio that are similar to asbestos fibers, which are a known cause of fibrosis and mesothelioma in humans. In fact, it is surprising to some that CNTs are being explored as a drug delivery platform based on similarities of CNTs to asbestos [8, 9]. There is now substantial evidence that CNTs can cause fibrogenic reactions in the lungs of mice and rats. Some of the first studies with CNTs demonstrated that they cause pulmonary inflammation and fibrosis when administered to the lung by intratracheal instillation or pharyngeal aspiration [6, 10–13]. These studies showed that CNTs, which aggregate into micron-sized bundles in aqueous media, stimulated the formation of inflammatory foci known as granulomas and fibrotic reactions within the lung parenchyma. Pulmonary fibrosis and/or granulomas were reported in mice or rats within days after the intratracheal instillation of SWCNTs [12, 13] or MWCNTs [11]. Moreover, SWCNTs delivered to the lung by instillation increased the production of platelet-derived growth factor (PDGF) and transforming growth factor (TGF)-β1, two growth factors that mediate the pathogenesis of fibrosis [6, 12]. Granulomas are associated with large CNT aggregates in the lung that are easily visible by light microscopy, while interstitial pulmonary fibrosis appears to be associated with dispersed CNTs that are detected by electron microscopy. While instillation or aspiration experiments demonstrated that CNTs are fibrogenic, a major caveat of early studies to assess toxicity was the fact that CNTs tended to aggregate into micron-sized particles. Obviously, such aggregates would not be suitable for drug delivery purposes, and a variety of modifications have been introduced to increase the dispersion of CNTs in biological media. Improved methods have been introduced for studying the potential health effects of CNTs, including the use of inhalation studies in rodents to provide a "real-world" exposure scenario and the use of better dispersants for preventing CNT aggregation. A study that compared instillation with inhalation indicated that MWCNTs inhaled by mice showed a more diffused pattern of deposition with less severe pathological lesions than those that were observed after intratracheal instillation of a bolus dose [14]. A recent study showed that SWCNTs dispersed with organic solvents to disrupt Van der Waals forces resulted in less deposition in the large airways but more in the interstitium, where the fibrotic response was significantly increased [15]. Therefore, increasing the dispersability of CNTs for drug delivery might not necessarily reduce toxicity but might simply alter the site of toxicity. If CNTs are to be used in drug delivery, two major obstacles that are likely to determine toxicity must be overcome. First, the length of CNTs must not exceed several micrometers as fibers longer than 10 to 15 micrometers (the approximate width of an alveolar macrophage) are difficult to clear from lung tissues via macrophage-mediated mechanisms. Second, the composition of CNTs must be carefully considered. Metals such as nickel, cobalt, and iron are commonly used as catalysts in the manufacture of CNTs and can be present as impurities in the final CNT product. These metals are known to cause pulmonary diseases in humans, including pulmonary fibrosis and asthma. Thus, if CNTs are ever to be used for drug delivery, an intelligent design must include a thorough consideration and elimination of these important toxicologic issues.

9.4 CNTs and Preexisting Allergic Asthma

Individuals with preexisting respiratory diseases such as asthma, bronchitis, and COPD represent the most susceptible populations to NP exposure [7]. If CNTs are to be used for drug delivery, careful consideration should be given to the possibility of exacerbation of preexisting disease, outweighing any beneficial effects. Several unique characteristics of diseased populations need to be taken into account when considering the use of CNTs in drug delivery. First, altered airway flow in patients with obstructive lung disease may increase airway deposition and relative concentration of inhaled NPs in these subjects compared to healthy individuals [16]. Second, the physicochemical characteristics of engineered CNTs may be affected by the presence in diseased airways of mucus, cellular debris, and other proteins, which may promote aggregation and change the surface area as well as charge. Third, the presence of activated inflammatory cells and structural airway cells may alter the response to an otherwise innocuous NP. Recent evidence by Ryman-Rasmussen *et al.* showed that MWCNTs caused little adverse pulmonary effect when delivered to the lungs of mice by inhalation except when the mice were first challenged with an allergen (ovalbumin) [17]. In this study, airway fibrosis was prominently increased by the combination of ovalbumin and MWCNTs whereas ovalbumin or MWCNTs alone caused only marginal increases in airway fibrosis (Fig. 9.2). Inhaled

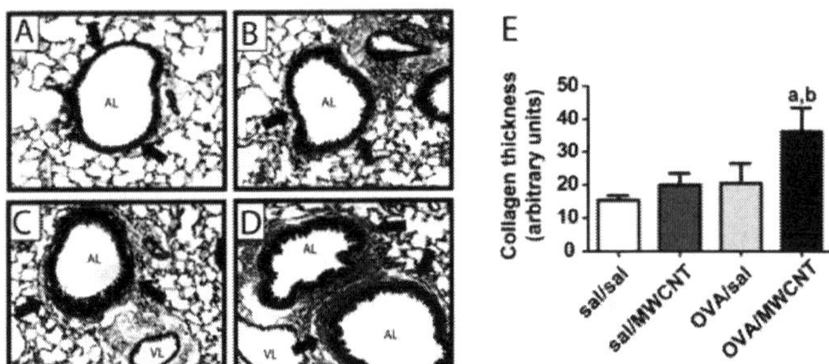

Figure 9.2 Exacerbation of allergen-induced airway fibrosis in mice by inhaled MWCNTs after two weeks. Panels A–D show cross-sections of airway from mice exposed to A) intranasal saline, then nebulized inhaled saline as a control (sal/sal), B) intranasal saline then exposed to inhaled MWCNTs (sal/MWCNT), C) intranasal ovalbumin (OVA) allergen followed by saline aerosol exposure (OVA/sal), and D) OVA intranasal preexposure followed by MWCNT inhalation exposure (OVA/MWCNT). Panel E shows the results of quantitative morphometry of airway fibrosis, where "a" denotes the significant difference between OVA/MWCNT and sal/MWCNT and "b" denotes the significant difference between OVA/MWCNT and OVA/sal [17]. See also Color Insert.

MWCNTs caused a significant increase in PDGF whereas ovalbumin significantly increased levels of TGF-β1 in bronchoalveolar lavage (BAL) fluid, and it was suggested that the combination of these two growth factors acted coordinately to stimulate fibroblast growth and collagen production by fibroblasts. In addition, IL-5 in BAL fluid was synergistically increased by the combination of ovalbumin and MWCNTs. The implications of this study were that CNTs pose a hazard to individuals with allergic lung inflammatory diseases such as asthma. Inoue and colleagues reported that MWCNTs administered intratracheally significantly increased ovalbumin-induced T-lymphocyte proliferation and amplified lung Th2 cytokines and chemokines compared to ovalbumin exposure alone [18]. More recent work by these investigators suggested that SWCNTs also can exacerbate murine allergic airway inflammation via the enhanced activation of Th2 immunity and increased oxidative stress and that exacerbation may be partly through the inappropriate activation of antigen-presenting cells, including dendritic cells [19]. Work by Park *et al.* showed that MWCNTs delivered to the lungs of mice may induce allergic responses through B cell activation and production of immunoglobulin E (IgE) in the absence of any allergen preexposure [20]. In addition to increased numbers of B cells, granuloma formation in the lung tissue and IgE production were also observed by MWCNTs in a dose-dependent manner. Nygaard *et al.* showed that SWCNTs as well as MWCNTs promote ovalbumin-induced allergic immune responses in the lungs of mice [21]. Collectively, the cumulative data from studies with mice from several different laboratories indicate that CNTs promote allergen-induced airway inflammation. Furthermore, CNTs alone might serve as allergens upon repeated exposures.

It is unknown if CNTs will cause or exacerbate asthma in humans. However, there is a correlation between particulate air pollution (especially the ultrafine, or nanoscale, fraction) and the incidence and severity of asthma attacks [22]. The same animal models of asthma that have been used to identify pulmonary effects of ultrafine particles that have some support from epidemiological data also identify immune responses and airway remodeling effects of MWCNTs that are relevant to asthma. Moreover, pulmonary exposure to nanometals exacerbated the asthmatic response in a mouse model [23]. Together, this information indicates that MWCNTs may cause or exacerbate human asthma. If true, then CNTs are probably poor candidates for delivering drugs aimed at the treatment of asthma. However, other nanomaterials, particularly fullerenes, have been shown to inhibit allergic airway inflammation in mice [24] and should be explored further with regard to drug delivery for the treatment of asthma.

9.5 CNTs and LPS-Induced Airway Inflammation

Bacterial lipopolysaccharide (LPS), a component of gram-negative bacteria cell walls, is a potent proinflammatory agent. LPS is ubiquitous in the environment and has been implicated in a number of occupational and environmental lung diseases in humans, including bronchitis, COPD, and asthma. Work by Inoue *et al.* showed that SWCNTs or MWCNTs moderately exacerbated lung inflammation, pulmonary vascular permeability, and lung expression of proinflammatory cytokines induced by LPS [25]. More recent work by Cesta *et al.* showed that LPS preexposure could exacerbate the fibrogenic potential of CNTs when delivered to the lungs of rats [26]. In this study, LPS alone did not cause fibrosis when delivered by intranasal aspiration but enhanced MWCNT-induced fibrosis and synergistically enhanced the production of macrophage and epithelial-derived PDGF, a mitogen and chemoattractant for lung fibroblasts. Companion *in vitro* experiments in lung fibroblasts were conducted to further investigate the mechanism of this effect. These studies confirmed that PDGF expression is up-regulated in lung fibroblasts after CNT exposure and that LPS exposure up-regulated expression of the PDGF-α receptor. LPS likely primes fibroblasts for heightened proliferative and migratory responses to PDGF by increasing levels of PDGF cell-surface receptors. This could be a mechanism for enhanced pulmonary fibrosis in rats exposed to LPS and CNTs. While not shown in these studies, it is also conceivable that LPS could adsorb to the surface of CNTs and be targeted to the distal airways and alveolar region of the lung upon inhalation exposure.

9.6 Effects of CNTs on Other Organ Systems

CNTs delivered to lungs by inhalation or aspiration can have adverse effects on distant organ systems, including the spleen and the heart. Mitchell and coworkers showed that inhaled MWCNTs do not cause significant interstitial lung inflammation or fibrosis in mice but do cause systemic immunosuppression and splenic oxidative stress [27]. The mechanism of splenic immunosuppression involves the release of TGF-β1 from the lungs of MWCNT-exposed mice, which enters the bloodstream to signal COX-2-mediated increases in prostaglandin-E_2 and IL-10 in the spleen, both of which play a role in suppressing T cell proliferation [28]. This result greatly differed from earlier work that showed that MWCNTs administered by intratracheal instillation resulted in pulmonary inflammation and fibrosis. Therefore, MWCNTs appear to cause less interstitial fibrosis when delivered by inhalation but significantly more injury, inflammation, and fibrosis when administered to the lungs by intratracheal or pharyngeal routes. However, even relatively low concentrations of inhaled MWCNTs appear to have significant effects on the systemic immune system. Erdely and coworkers reported that SWCNTs or MWCNTs deposited into the lung induced acute lung and systemic effects, suggesting that the systemic response, if chronic and persistent, could trigger or exacerbate cardiovascular dysfunction and disease [29].

9.7 DNA Damage and Aneuploidy Caused by CNTs

Several studies have shown that CNTs have the potential to cause DNA damage in lung cell types. For example, work by Sargent *et al.* showed that SWCNTs caused fragmented centrosomes, multiple mitotic spindle poles, anaphase bridges, and aneuploid chromosome number in cultured primary or immortalized human airway epithelial cell types [30]. This work was the first to show the disruption of the mitotic spindle by SWCNTs, and the authors noted that the similar size and geometry of SWCNT bundles might account for the physical interaction of SWCNTs with the microtubules that form the mitotic spindle. Zhu *et al.* assessed the DNA damage response to MWCNTs in mouse embryonic stem cells and found that MWCNTs can accumulate and induce apoptosis in these cells and activate the tumor suppressor protein p53, a marker of DNA damage [31]. Collectively, these studies show that SWCNTs and MWCNTs can exert genotoxic effects in certain *in vitro* systems but do not show that CNTs can translocate to sensitive cells *in vivo* to cause genotoxic effects. Evidence of this was provided by Patlolla *et al.*, who showed that repeated-dose (i.p.) administration of nonfunctionalized and COOH-functionalized MWCNTs to Swiss-Webster mice caused dose-dependent genotoxic effects of structural chromosome aberrations, DNA damage, and micronuclei formulation [32]. A dose-dependent decrease in mitotic index was also observed. The incidence of genotoxic effects was apparently higher for functionalized MWCNTs. Given the important role that genotoxicity plays in the development of cancer, evidence of genotoxicity of CNTs in screening-level assays, which are commonly conducted during the development of therapeutic agents and consumer products, may prove to be particularly limiting to further development or marketing of CNT-containing products.

9.8 Pleural Toxicity of CNTs

As mentioned earlier in this chapter, CNTs are cleared from the lungs via macrophage-mediated mechanisms and one route that macrophages take to remove particles from the lungs is across the pleural lining via the lymphatic drainage. Therefore, CNTs have the opportunity to interact with the mesothelial lining that makes up the pleura. The durable nature of CNTs along with their fiberlike shape could pose a problem of long-term lung persistence, which might result in asbestos-like behavior and carcinogencity (i.e., mesothelioma). Studies by Poland and colleagues showed that injection of long MWCNTs into the peritoneal cavity of mice (a surrogate for the pleura) induced inflammation, suggesting that MWCNTs have asbestos-like pathogenicity [33]. This study was important because it demonstrated that long MWCNTs, but not short MWCNTs, caused granuloma formation at the pleura similar to that caused by long asbestos fibers. More relevant inhalation studies in mice by Ryman-Rasmussen *et al.* show that MWCNTs reach the pleura (the anatomic site of mesothelioma) and cause significant increases in pleural fibrosis [34]. However, no indication of mesothelioma was found in mice that inhaled MWCNTs. Therefore, it is unknown whether CNTs represent a new cancer risk factor similar to asbestos. While CNTs cause lung fibrotic reactions in the interstitium and pleura of mice, their carcinogenic potential has not been adequately addressed. Longer term, low dose studies with CNTs will need to be undertaken to determine possible carcinogenicity. Whether

CNTs cause mesothelioma is obviously a major issue that must be thoroughly addressed if CNTs are to be widely used in consumer products or as drug delivery platforms.

9.9 Conclusions

Engineered nanomaterials, including CNTs, represent novel drug delivery platforms. However, the potential risks of CNTs should be carefully considered when designing nanoscale drug delivery platforms. Because of the infancy of the field of nanotechnology, there is no epidemiologic data to indicate health hazards for the majority of nanomaterials, including those that could be used for drug delivery. CNTs may contain nanosized metals (e.g., nickel, cobalt, and vanadium) that are present as residual catalysts; such CNTs may represent risk factors for lung diseases as many of these metals in their native form are known to have fibrogenic or carcinogenic effects in humans. Moreover, some CNTs have shown inflammatory, immune, and/or fibrogenic effects in the lungs of mice and rats, indicating that caution should be taken in the handling of these materials. The development of nanotechnology-based therapeutics, including those based on CNTs, will be an ongoing challenge due to the anticipated evolution of nano-specific regulatory guidelines and safety-testing protocols. However, this challenge is expected to be overcome as experience is gained and quality physicochemical and toxicity data becomes more widely available, thus leading to the development of effective nanotechnology-based therapies for a range of respiratory diseases.

References

1. De Jong, W.H., and Borm, P.H. (2008) Drug delivery and nanoparticles: applications and hazards. *Int. J. Nanomed.* **3**(2), pp. 133–149.
2. Card, J.W., Zeldin, D.C., Bonner, J.C., and Nestmann, E.R. (2008) Pulmonary applications and toxicity of engineered nanoparticles. *Am. J. Physiol.: Lung Cell Mol. Physiol.* **295**(3), pp. L400–L411.
3. Bonner, J.C., Card, J.W., and Zeldin, D.C. (2009) Nanoparticle-mediated drug delivery and pulmonary hypertension. *Hypertension.* **53**, pp. 751–753.

4. Prato, M., Kostarelos, K., and Bianco, A. (2008) Functionalized carbon nanotubes in drug design and discovery. *Acc. Chem. Res.* **41**(1), pp. 60–68.

5. Konduru, N.V., Tyurina, Y.Y., Feng, W., Basova, L.V., Belikova, N.A., Bayir, H., Clark, K., Rubin, M., Stolz, D., Vallhov, H., Scheynius, A., Witasp, E., Fadeel, B., Kichambare, P.D., Star, A., Kisin, E.R., Murray, A.R., Shvedova, A.A., and Kagan, V.E. (2009) Phosphatidylserine targets single-walled carbon nanotubes to professional phagocytes in vitro and in vivo. *PLoS One*. **4**(2), p. e4398.

6. Mangum, J.B., Turpin, E.A., Antao-Menezes, A., Cesta, M.F., Bermudez, E., and Bonner, J.C. (2006) Single-walled carbon nanotube (SWCNT)-induced interstitial fibrosis in the lungs of rats is associated with increased levels of PDGF mRNA and the formation of unique intercellular carbon structures that bridge alveolar macrophages in situ. *Part Fibre Toxicol.* **3**, p. 15.

7. Bonner, J.C. (2010) Nanoparticles as a potential cause of pleural and interstitial lung disease. *Proc. Am. Thorac. Soc.* **7**(2), pp. 138–141.

8. Borm, P.J., Robbins, D., Haubold, S., Kuhlbusch, T., Fissan, H., Donaldson, K., Schins, R., Stone, V., Kreyling, W., Lademann, J., Krutmann, J., Warheit, D., and Oberdorster, E. (2006) The potential risks of nanomaterials: a review carried out for ECETOC. *Part Fibre Toxicol.* **3**, p. 11.

9. Donaldson, K., Aitken, R., Tran, L., Stone, V., Duffin, R., Forrest, G., and Alexander, A. (2006) Carbon nanotubes: a review of their properties in relation to pulmonary toxicology and workplace safety. *Toxicol. Sci.* **92**(1), pp. 5–22.

10. Lam, C.W., James, J.T., McCluskey, R., and Hunter, R.L. (2004) Pulmonary toxicity of single-wall carbon nanotubes in mice 7 and 90 days after intratracheal instillation. *Toxicol. Sci.* **77**, pp. 126–134.

11. Muller, J., Huaux, F., Moreau, N., Misson, P., Heilier, J.F., Delos, M., Arras, M., Fonseca, A., Nagy, J.B., and Lison, D. (2005) Respiratory toxicity of multi-wall carbon nanotubes. *Toxicol. Appl. Pharmacol.* **207**, pp. 221–231.

12. Shvedova, A.A., Kisin, E.R., Mercer, R., Murray, A.R., Johnson, V.J., Potapovich, A.I., Tyurina, Y.Y., Gorelik, O., Arepalli, S., Schwegler-Berry, D., Hubbs, A.F., Antonini, J., Evans, D.E., Ku, B.K., Ramsey, D., Maynard, A., Kagan, V.E., Castranova, V., and Baron, P. (2005) Unusual inflammatory and fibrogenic pulmonary responses to single-walled carbon nanotubes in mice. *Am. J. Physiol. Lung Cell Mol. Physiol.* **289**, pp. L698–L708.

13. Warheit, D.B., Laurence, B.R., Reed, K.L., Roach, D.H., Reynolds, G.A., and Webb, T.R. (2004) Comparative pulmonary toxicity assessment of single-wall carbon nanotubes in rats. *Toxicol. Sci.* **77**, pp. 117–125.

14. Li, J.G., Li, W.X., Xu, J.Y., Cai, X.Q., Liu, R.L., Li, Y.J., Zhao, Q.F., and Li, Q.N. (2007) Comparative study of pathological lesions induced by multiwalled carbon nanotubes in lungs of mice by intratracheal instillation and inhalation. *Environ. Toxicol.* **22**, pp. 415–421.

15. Mercer, R.R., Scabilloni, J., Wang, L., Kisin, E., Murray, A.R., Schwegler-Berry, D., Shvedova, A.A., and Castranova, V. (2008) Alteration of deposition pattern and pulmonary response as a result of improved dispersion of aspirated single-walled carbon nanotubes in a mouse model. *Am. J. Physiol. Lung Cell Mol. Physiol.* **294**, pp. L87–L97.

16. Anderson, P.J., Wilson, J.D., and Hiller, F.C. (1990) Respiratory tract deposition of ultrafine particles in subjects with obstructive or restrictive lung disease. *Chest.* **97**(5), pp. 1115–1120.

17. Ryman-Rasmussen, J.P., Tewksbury, E.W., Moss, O.R., Cesta, M.F., Wong, B.A., and Bonner, J.C. (2009) Inhaled multiwalled carbon nanotubes potentiate airway fibrosis in a murine model of allergic asthma. *Am. J. Resp. Cell Mol. Biol.* **40**(3), pp. 349–358.

18. Inoue, K., Koike, E., Yanagisawa, R., Hirano, S., Nishikawa, M., and Takano, H. (2009) Effects of multi-walled carbon nanotubes on a murine allergic airway inflammation model. *Toxicol. Appl. Pharmacol.* **237**(3), pp. 306–316.

19. Inoue, K., Yanagisawa, R., Koike, E., Nishikawa, M., and Takano, H. (2010) Repeated pulmonary exposure to single-walled carbon nanotubes exacerbates allergic inflammation of the airway: Possible role of oxidative stress. *Free Radic. Biol. Med.* **48**(7), pp. 924–934.

20. Park, E.J., Cho, W.S., Jeong, J., Yi, J., Choi, K., and Park, K. (2009) Pro-inflammatory and potential allergic responses resulting from B cell activation in mice treated with multi-walled carbon nanotubes by intratracheal instillation. *Toxicology.* **259**(3), pp. 113–121.

21. Nygaard, U.C., Hansen, J.S., Samuelsen, M., Alberg, T., Marioara, C.D., and Løvik, M. (2009) Single-walled and multi-walled carbon nanotubes promote allergic immune responses in mice. *Toxicol. Sci.* **109**(1), pp. 113–123.

22. Gavett, S.H., and Koren, H.S. (2001) The role of particulate matter in exacerbation of atopic asthma. *Int. Arch. Allergy Immunol.* **124**(1–3), pp. 109–112.

23. Hussain, S., VAnoirbeek, J.A., Luyts, K., De Vooght, V., Verbeken, E., Thomassen, L.C., Martens, J.A., Dinsdale, D., Boland, S., Marano, F., Nemery, B., and Hoet, P.H. (2010) Lung exposure to nanoparticles modulates an asthmatic response in a mouse model of asthma. *Eur. Respir. J.* [Epub ahead of print].

24. Ryan, J.J., Bateman, H.R., Stover, A., Gomez, G., Norton, S.K., Zhao, W., Schwartz, L.B., Lenk, R., and Kepley, C.L. (2007) Fullerene nanomaterials inhibit the allergic response. *J. Immunol.* **179**, pp. 665–672.

25. Inoue, K., Takano, H., Koike, E., Yanagisawa, R., Sakurai, M., Tasaka, S., Ishizaka, A., and Shimada, A. (2008) Effects of pulmonary exposure to carbon nanotubes on lung and systemic inflammation with coagulatory disturbance induced by lipopolysaccharide in mice. *Exp. Biol. Med.* **233**(12), pp. 1583–1590.

26. Cesta, M.F., Ryman-Rasmussen, J.P., Wallace, D.G., Masinde, T., Hurlburt, G., Taylor, A.J., and Bonner, J.C. (2010) Bacterial lipopolysaccharide enhances PDGF signaling and pulmonary fibrosis in rats exposed to carbon nanotubes. *Am. J. Respir. Cell Mol. Biol.* **43**(2), pp. 142–151.

27. Mitchell, L.A., Gao, J., Wal, R.V., Gigliotti, A., Burchiel, S.W, and McDonald, J.D. (2007) Pulmonary and systemic immune response to inhaled multiwalled carbon nanotubes. *Toxicol. Sci.* **100**, pp. 203–214.

28. Mitchell, L.A., Lauer, F.T., Burchiel, S.W., and McDonald, J.D. (2009) Mechanisms for how inhaled multiwalled carbon nanotubes suppress systemic immune function in mice. *Nature Nanotech.* **4**(7), pp. 451–456.

29. Erdely, A., Hulderman, T., Salmen, R., Liston, A., Zeidler-Erdely, P.C., Schwegler-Berry, D., Castranova, V., Koyama, S., Kim, Y.A., Endo, M., and Simeonova, P.P. (2009) Cross-talk between lung and systemic circulation during carbon nanotube respiratory exposure. Potential biomarkers. *Nano Lett.* **9**(1), pp. 36–43.

30. Sargent, L.M., Shvedova, A.A., Hubbs, A.F., Salisbury, J.L., Benkovic, S.A., Kashon, M.L., Lowry, D.T., Murray, A.R., Kisin, E.R., Friend, S., McKinstry, K.T., Battelli, L., and Reynolds, S.H. (2009) Induction of aneuploidy by single-walled carbon nanotubes. *Environ. Mol. Mutagen.* **50**(8), pp. 708–717.

31. Zhu, L., Chang, D.W., Dai, L., and Hong, Y. (2007) DNA damage induced by multiwalled carbon nanotubes in mouse embryonic stem cells. *Nano Lett.* **7**(12), pp. 3592–3597.

32. Patlolla, A.K., Hussain, S.M., Schlager, J.J., Patlolla, S., and Tchounwou, P.B. (2010) Comparative study of the clastogenicity of functionalized and non-functionalized multi-walled carbon nanotubes in

bone marrow cells of Swiss-Webster mice. *Environ. Toxicol.* **25**(6), pp. 608–621.

33. Poland, C.A., Duffin, R., Kinloch, I., Maynard, A., Wallace, W.A.H., Seaton, A., Stone, V., Brown, S., MacNee, W., and Donaldson, K. (2008) Carbon nanotubes introduced into the abdominal cavity of mice show asbestos-like pathogenicity in a pilot study. *Nat. Nanotechnol.* **3**(7), pp. 423–428.
34. Ryman-Rasmussen, J.P., Cesta, M.F., Brody, A.R., Shipley-Phillips, J.K., Everitt, J., Tewksbury, E.W., Moss, O.R., Wong, B.A., Dodd, D.E., Andersen, M.E., and Bonner, J.C. (2009) Inhaled multi-walled carbon nanotubes reach the subpleural tissue in mice. Nat. Nanotechnol. **4**(11), pp. 747–751.

Index

Color Insert

Chapter 1

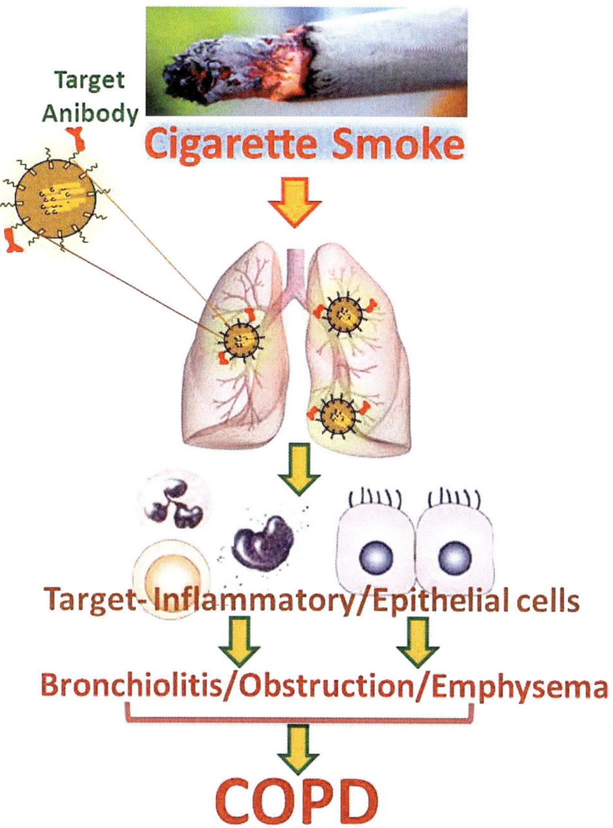

Figure 1.1

Chapter 2

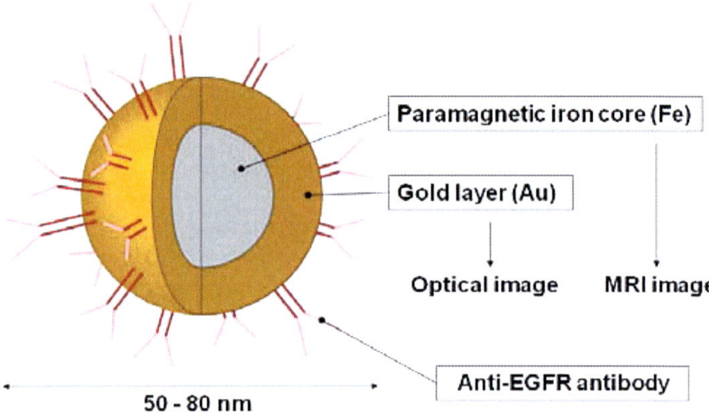

Figure 2.1

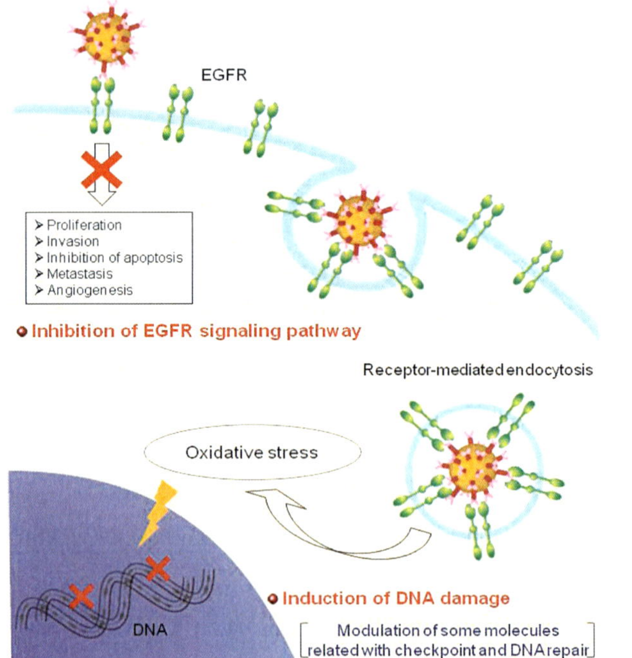

Figure 2.2

Chapter 4

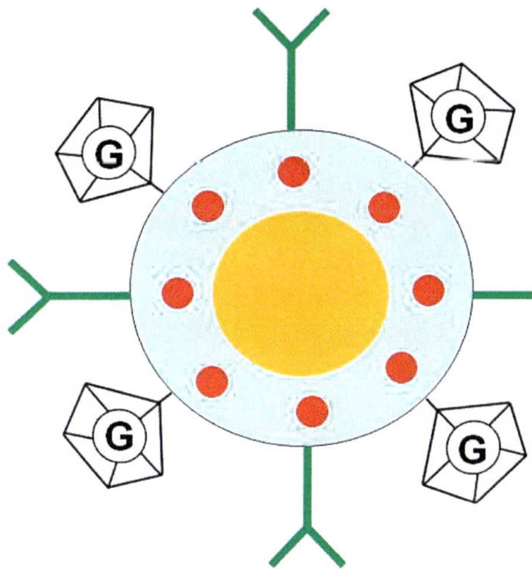

Figure 4.1

Chapter 5

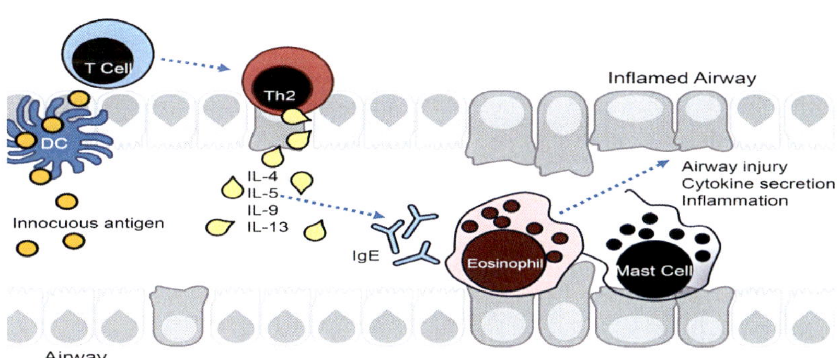

Figure 5.1

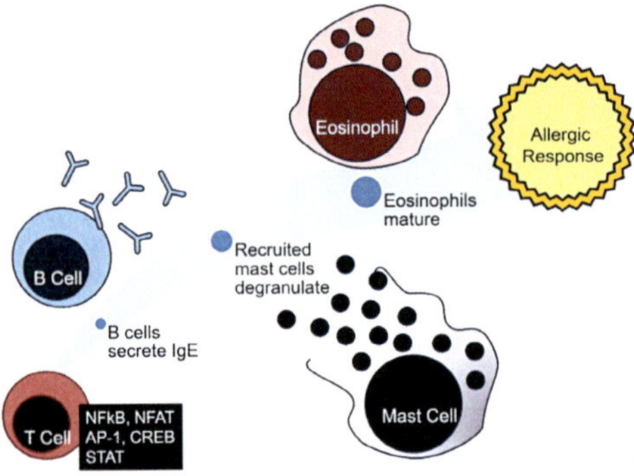

Figure 5.2

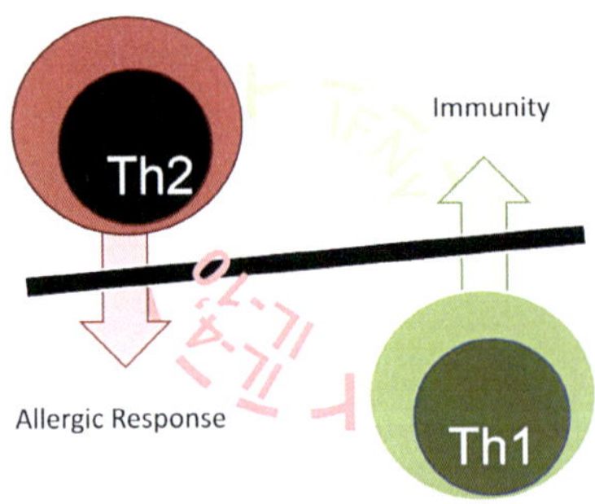

Figure 5.3

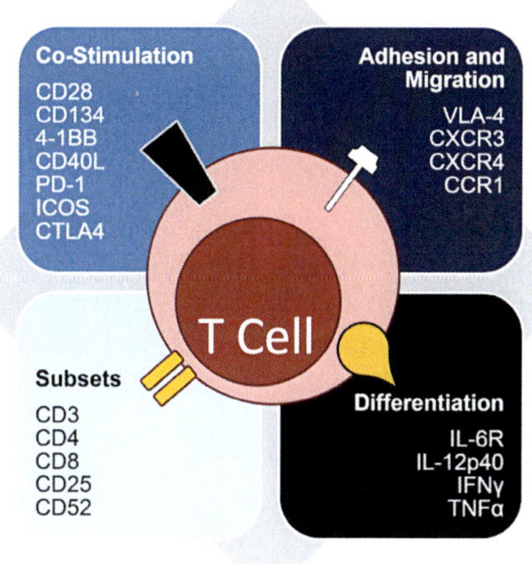

Figure 5.4

Chapter 6

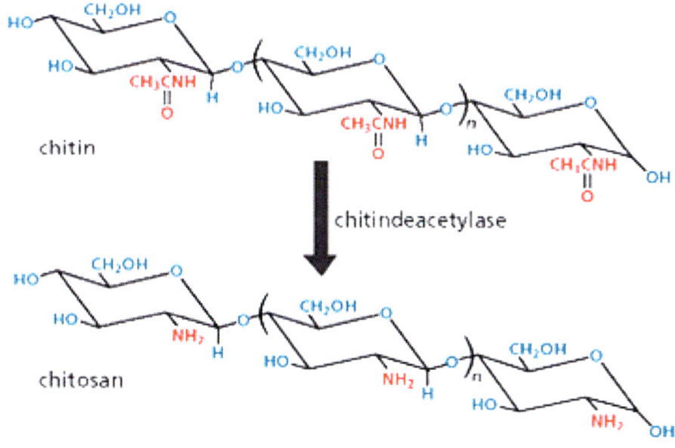

Figure 6.1

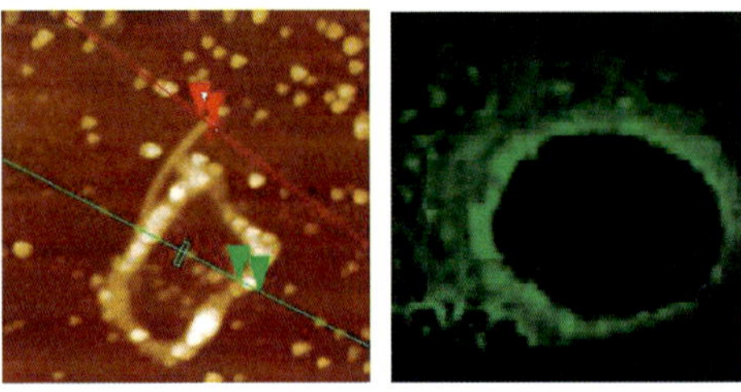

Figure 6.2

Chapter 8

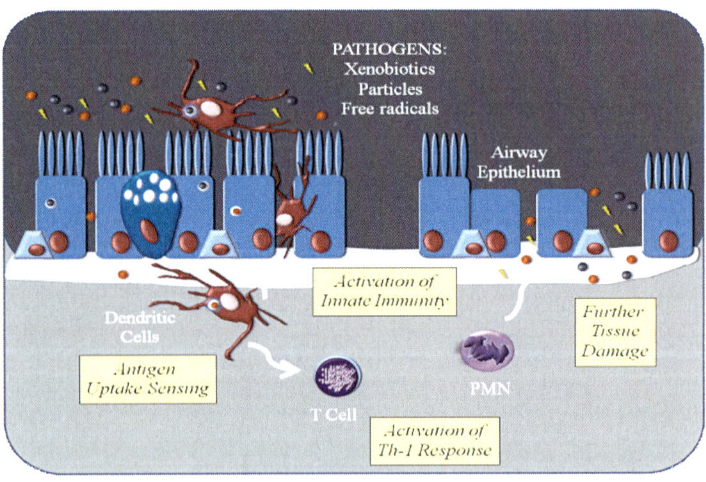

Figure 8.1

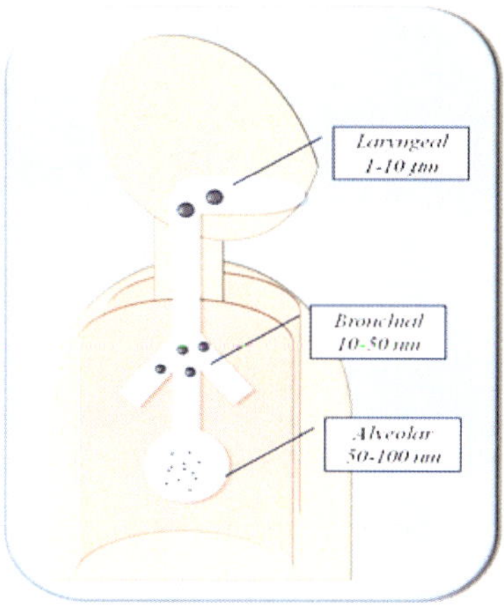

Figure 8.2

Chapter 9

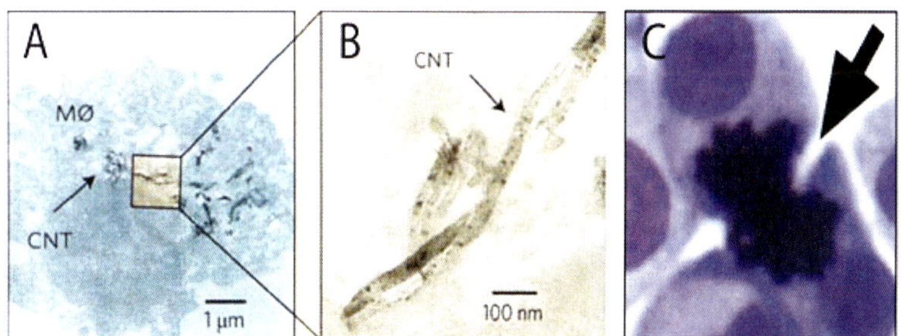

Figure 9.1

A
B
C
D
AL
VL
E
Collagen thickness (arbitrary units)
50
40
30
20
10
0
a,b
sal/sal
sal/MWCNT
OVA/sal
OVA/MWCNT

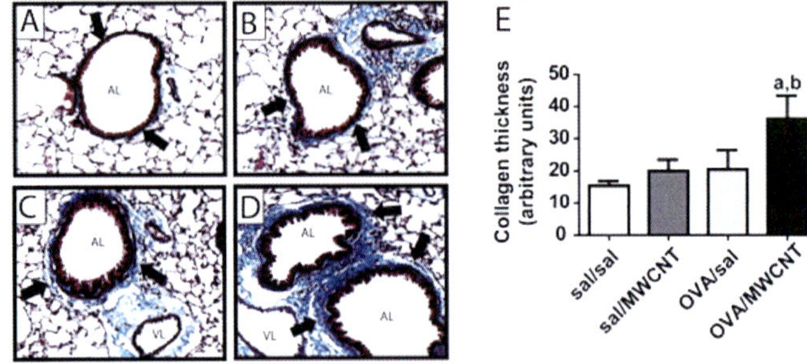

Figure 9.2